# Materials Characterization

Ramiro Pérez Campos
Antonio Contreras Cuevas
Rodrigo Esparza Muñoz
Editors

# Materials Characterization

Springer

*Editors*
Ramiro Pérez Campos
Centro de Física Aplicada y Tecnología Avanzada of Universidad Nacional Autónoma de México (UNAM)
Querétaro, Querétaro, México

Rodrigo Esparza Muñoz
Centro de Física Aplicada y Tecnología Avanzada of Universidad Nacional Autónoma de México (UNAM)
Querétaro, México

Antonio Contreras Cuevas
Instituto Mexicano del Petróleo (IMP)
San Bartolo Atepehuacan, México

ISBN 978-3-319-38406-1 ISBN 978-3-319-15204-2 (eBook)
DOI 10.1007/978-3-319-15204-2

Springer Cham Heidelberg New York Dordrecht London

Softcover reprint of the hardcover 1st edition 2015


Printed on acid-free paper

Springer International Publishing AG Switzerland is part of Springer Science+Business Media (www.springer.com)

# Preface

This book entitled Materials Characterization includes a compilation of the best research works presented in the Symposium 5B "Structural and chemical characterization of metals, alloys and compounds," by XXIII International Materials Research Congress (IMRC-2014). The XXIII International Materials Research Congress was held on August 17–21, 2014, in Cancún, México. It was organized by the Sociedad Mexicana de Materiales (SMM). About 1,400 specialized scientists from more than 40 countries participated in the 30 different symposia, workshops, plenary lectures, and tutorial courses. The 30 symposia that comprise the technical program of IMRC 2014 are grouped in several clusters, namely Nanoscience and Nanotechnology, Biomaterials, Materials for Energy, Fundamentals of Materials Science, Materials Characterization, Materials for Specific Applications, Magnetic and Electronic Materials, and General.

The chapters contained in this book cover state-of-the-art and experimental techniques commonly used in modern materials characterization. In addition, the book introduces advanced techniques, including Scanning Probe Microscopy, Aberration-Corrected Microscopy, Raman Spectroscopy, Fourier Transform Infrared Spectroscopy, Field Emission Scanning Electron Microscopy (FESEM), among others.

This book covers several aspects of the structural and chemical characterization of the materials in the following areas: metals, alloys, steels, composites, welding, nanomaterials, and surface coatings, among others. They are amorphous, crystalline, powders, coatings, fibers, thin films, and so forth, which were prepared with different techniques. The structural characterization techniques include scanning electron microscopy (SEM), X-ray diffraction (XRD), transmission electron microscopy (TEM), aberration correction by scanning transmission electron microscopy (STEM), Raman spectroscopy, optical microscopy (OM), Fourier transform infrared spectroscopy (FTIR), differential thermal analysis (DTA), differential scanning calorimetry (DSC), thermogravimetry analysis (TGA), thermoluminescence (TL), laser emission, and so forth. Theoretical models from these properties are included too.

This specialized book on materials characterization contains 21 chapters based on contributions presented at the symposium 5B. All manuscripts included in this special issue have been accepted after peer review. The structure of the book is classified into eight parts: Characterization of Welding and Joining of Materials, Characterization of Nanostructured Materials, Characterization of Steels Used in the Oil Industry, Characterization of Stainless Steels, Characterization of Composite Materials, Characterization of Materials for Medical Applications, Characterization of Materials for Industrial Applications, and Characterization of Intermetallic Materials.

Querétaro, México — Ramiro Pérez Campos
Distrito Federal, México — Antonio Contreras Cuevas
Querétaro, México — Rodrigo Esparza Muñoz

# Acknowledgments

We would like to thank Springer editorial for the invitation and kind acceptance to publish this book as well as the committee assigned for the edition of the book and the reviewers of the chapters for their valuable comments, which have certainly helped to improve the quality of the manuscripts. We also wish to thank the Sociedad Mexicana de Materiales (SMM), Centro de Física Aplicada y Tecnología Avanzada (CFATA) from Universidad Nacional Autonoma de México (UNAM), and Instituto Mexicano del Petróleo (IMP) for their support in organizing the Symposium 5B "Structural and chemical characterization of metals, alloys and compounds," by XXIII International Materials Research Congress (IMRC-2014) from which are the chapters included in this book. Also, we would like to thank all the authors of the chapters who have worked to improve the quality of the chapters, which are result of their valuable research work; and make possible the edition of this book.

# Contents

# Biography of Editors

**Dr. Ramiro Pérez Campos** obtained his Ph. D. in Physics at the University of Alberta, Canada, in 1983. He also carried out his postdoctoral work at the KFA in Jülich, Germany. He was awarded an Alexander Von Humboldt postdoctoral scholarship. He is the author of more than 170 scientific papers, all of them related to the applications of the transmission electron microscope in materials science. He is the current Director of Centro de Física Aplicada y Tecnología Avanzada of the Universidad Nacional Autónoma de México.

**Dr. Antonio Contreras Cuevas** is a pipeline materials research specialist for Mexican Petroleum Institute since 2003. He obtained a master's degree in 1997 and Ph.D. in 2002 in Metallurgy and Materials Science from Universidad Nacional Autónoma de México (UNAM). He is a member of NACE and Mexican Materials Society, and he belongs to National Research System. The main research areas include stress corrosion cracking (SCC), root cause analysis (RCA), materials characterization, and pipeline integrity. He is the author of more than 50 scientific articles related to materials characterization by different techniques. Actually, he is a researcher of Mexican Petroleum Institute in México.

**Dr. Rodrigo Esparza Muñoz** obtained his Ph.D. in Metallurgy and Materials Science at the Universidad Michoacana de San Nicolás de Hidalgo (UMSNH), México, in 2007. He carried out two postdoctoral stays, one of them in the Universidad Nacional Autónoma de México (UNAM) and the other one in the University of Texas at San Antonio (UTSA). He is the author of more than 50 scientific papers, most of them related to the characterization of nanomaterials using high-resolution transmission electron microscopy and aberration-corrected scanning transmission electron microscopy. Actually, he is a researcher of Centro de Física Aplicada y Tecnología Avanzada from Universidad Nacional Autónoma de México.

# Part I
# Characterization of Welding and Joining of Materials

The first part of this book includes three chapters related with characterization welding materials by two joining process more employed on the industry and more specifically in automotive sector due to better quality in the welded joints are needed as well as higher volume in production. These techniques include gas metal arc welding (GMAW) and shielded metal arc welding (SMAW).

The material welded by GMAW technique includes EN S235JR low carbon structural steel, which is used in manufacturing components for new renewable energy sources. An important factor that has to be taken into account is the weldability behavior, since it is within a relevant field of structural engineering that requires the new technology of robotic welding. Different parameters used in the process are analyzed.

Shielded metal arc welding was used to weld an ASTM A36 steel in the T-welded connections repaired by grinding and wet welding. T-welded connections are employed and used to build fixed platforms in the Gulf of México. The main problems in T-welded connections are the porosity and cracking occurred in the weld pool, reflected in low mechanical properties, which is addressed in one of the chapters of this part.

Additionally, is included a chapter with interesting results about distortion in dissimilar stainless steels joints welded by GMAW process. Distortion became a potential problem in this process, since the joint undergoes significant thermal expansion and contraction, longitudinal and transverse shrinkages and leads to a non-desired welded structure.

The main techniques used in the characterization of these materials are:

- Optical microscopy (OM).
- Scanning electron microscopy (SEM).
- Energy-dispersive X-ray spectroscopy (EDS).
- Mechanical tests.
- Geomagic Design (3D CAD simulation).

# Chapter 1
# Metallurgical Investigation of Heat Input on Robot-Welded Joints Using GMAW Pulsed Process on Structural Steel

**I. Guzman-Flores, B. Vargas-Arista, J.J. Gasca-Dominguez, J. del Prado, E. Garfias-Garcia, and L. Rios-Vargas**

**Abstract** The effect of heat input on microstructural zones, hardness and fracture of T-joints in EN S235JR carbon steel was evaluated using the Vickers hardness, SEM, and fractography on impact fractured specimens. The gas metal arc welding (GMAW) process with pulsed metal arc transfer and different welding speeds was performed by an industrial robot using a teaching pendant. Microstructural observations revealed differences in the size and quantity of phases exhibited in the solidified weld bead and recrystallized heat affected zone (HAZ) as well as a refinement of ferrite with different morphologies when a greater heat input was generated. The hardness results indicated greater hardening with greater heat input for the weld bead than the HAZ, which was associated with fine secondary Widmastätten ferrite sideplates growing from columnar grains separated by a larger amount of fine acicular ferrite. At a speed of 35 cm/min, the greater heat input produced a greater increase in volume fraction and width of HAZ with complex microstructures formed by fine-grained ferrite and acicular ferrite. The weld bead reached a greater hardening and volume fraction compared to the HAZ linked to fine acicular ferrite. The hardening of weld bead was confirmed by fractography that showed the presence of a small microvoid morphology

---

I. Guzman-Flores • J.J. Gasca-Dominguez • J. del Prado • L. Rios-Vargas
Manufacturera de cigueñales de México (MACIMEX), Centro de Investigación Innovación y Desarrollo Avanzado, Km. 38.5 Carretera la Marquesa, Tenango del Valle, Estado de México 52300, México
e-mail: Isidro.Guzman@macimex.com; JuanJose.Gasca@macimex.com; Joaquin.DelPrado@macimex.com; Leopoldo.Rios@macimex.com

B. Vargas-Arista (✉)
División de Estudios de Posgrado e Investigación, Instituto Tecnológico de Tlalnepantla, Av. Instituto Tecnológico s/n, Col. La Comunidad, Tlalnepantla de Baz, Estado de México 54070, México
e-mail: benjaminvar@yahoo.com.mx

E. Garfias-Garcia
División de CBI, Departamento de Materiales, Área de Ingeniería de Materiales, UAM – Azcapotzalco Av. San Pablo No. 180, Col. Reynosa Tamaulipas, Azcapotzalco 02200, México
e-mail: D.F.elgg@correo.azc.uam.mx



R. Pérez Campos et al. (eds.), *Materials Characterization*,
DOI 10.1007/978-3-319-15204-2_1

caused by Charpy ductile failure with reduction of energy. The engineering and scientific relevance of this work is the original knowledge regarding the fractography and HAZ microstructural evolution of robotic welded joints as a function of heat input. The main problem solved with this research is the lack of knowledge regarding the welding metallurgy of robotic GMAW process.

**Keywords** Robotic welding • Fracture • HAZ • Hardening • Widmastätten ferrite

## 1.1 Introduction

When the construction of a welded solar structure is planned as a new renewable energy source, an important factor that has to be taken into account is the enhanced weldability behavior of the potential material to be selected, like hot rolled carbon steels for structural purpose. In particular, EN S235JR low carbon structural steel could be proposed for the manufacturing of these components, since it is within a relevant field of structural engineering that requires the new technology of robotic welding with pulsed arc. Gas metal arc welding (GMAW) for one pass joints is one of the most used processes in the automotive industry due to better quality in the welded joints as well as higher volume in production [1].

There has been a great amount of research done on automated welding using a teaching pendant, taking into account the design, integration, simulation, and field testing trials of new types of welding systems using axis manipulators, modularized controllers, and good mobility and without the use of any additional finishing by manual welders, resulting in functionality and adequate performance capabilities [2]. A key outcome of the GMAW process is the formation of heat affected zone (HAZ), which is made up of complex recrystallized microstructures classified as coarse and fine-grained zones. Fewer contributions have been published on the fractography and microstructural characterization of the HAZ and weld bead generated in robot-welded joints on carbon steels such as EN S235JR. Some of the research is related to the study of the effect of various parameters on penetration in 6842 steel welded by GMAW robotics. The current (95–115 A), arc voltage and welding speed (40–80 cm/min) were chosen as variables. The depths of penetration were measured for each specimen welded. Increasing current increased the depth of penetration. The highest penetration was observed at a speed of 60 cm/min [3]. Another example is found in the microstructural and mechanical characterizations performed on microstructural zones of multi-layered welded joints in ASTM A633 steel. The wire feed speed and voltage of the GMAW process was modified to analyze their effect. Both kinds of HAZ (conventional and reheated) showed hardening as a function of the increment in wire speed, resulting in higher current and net heat input, which favored a drastic reduction in Charpy impact energy. The weld bead showed a decrease in Vickers hardness and volumetric fraction as a function of the increase in speed, which was associated with the drastic reduction in the amount of columnar ferritic grains, Widmastätten and acicular ferrites induced by greater heat input [4].

This work focuses on the comparative analysis of the microstructure, microhardness, and fractography of impact specimens for weld bead and HAZ with extreme values of heat input in robot-welded joints of the aforementioned steel. The justification of this investigation is based on the fact that the HAZ is a detrimental hardened microstructural zone of lower toughness where cold cracking or hydrogen induced cracking and stress corrosion cracking are nucleated. Besides, the amount of coarse-grained HAZ must be minimized by correct parameters of the weld run sequence to avoid cracks after a welding cycle or unfavorable stress relief [5], which could drastically reduce the service life of welded joints. The originality of this study is found in the application of the welding metallurgy of the robotic welded joints considering the heat input.

## 1.2 Experimental

Specimens of the welded joint were obtained from plates of S235JR carbon steel with dimensions measuring 250 mm in length, 250 mm in width, and 12.7 mm in thickness which were welded longitudinally using a single-pass robotic GMAW process with a double T configuration under welding procedure specifications (WPSs) including an ER70S6 carbon steel electrode of 1.12 mm in diameter, a gas mixture of 82 % Ar/18 % $CO_2$ and a flow of 15 L/min, according to EN1011-2 [5] and ISO15609-1 [6]. The GMAW process using pulsed metal arc transfer with different welding speeds and heat inputs [1, 3] was performed by a Comau stationary robot with teaching pendant, a C5G arm with six motion axes, a work station with two working tables and transpuls 4,000 power machining.

The chemical composition given for the filler and base metals in as-received condition, in accordance with AWS A5.18 [7] and EN 10025-2 [8] respectively, is listed in Table 1.1. The values of carbon equivalent ($C_{eq}$) were calculated by Eq. (1.1) [9] which complied with EN 10025-2 [8]. $C_{eq}$ is a parameter to classify the weldability in three zones: easy to weld, weldable, and difficult to weld, according to Graville diagram [10]. In this case, these values were similar for both metals and compared to the Graville diagram it was seen that both metals were in the "easy to weld" zone, assuring the weldability of the welded joints without cracking after the welding process.

$$C_{eq} = C + \frac{Mn}{6} + \frac{Cr}{5} + \frac{Mo}{5} + \frac{V}{5} + \frac{Ni}{15} + \frac{Cu}{15} \tag{1.1}$$

**Table 1.1** Chemical composition of the ER70S6 filler metal and S235JR base metal

| | Element (wt%) | | | | | | | | | | |
|---|---|---|---|---|---|---|---|---|---|---|---|
| Material | C | Mn | Si | Cr | Ni | Mo | Nb | Cu | P | S | $C_{eq}$ |
| ER70S6 | 0.10 | 1.15 | 0.41 | 0.05 | 0.04 | 0.01 | 0.010 | 0.085 | 0.002 | 0.008 | 0.31 |
| S235JR | 0.11 | 0.86 | 0.22 | 0.12 | 0.14 | 0.04 | – | 0.399 | 0.036 | 0.035 | 0.32 |

**Table 1.2** Mechanical properties of the micro-wire and base metal

| Material | Vickers hardness HVN 500 $g_f$ | Yield strength, $\sigma_o$ (MPa) | Ultimate tensile strength, $\sigma_u$ (MPa) | Elongation (%) |
|---|---|---|---|---|
| ER70S-6 | 198.2 | 430 | 530 | 24 |
| S235JR | 162.7 | 371 | 519 | 31 |

**Table 1.3** Robotic welding parameters under study

| Specimen | Welding current (A) | Arc voltage (V) | Welding speed (cm/min) | Net heat input $Q_{net}$ (kJ/mm) |
|---|---|---|---|---|
| CM13 | 188 | 23 | 58 | 0.454 |
| CL1 | 238 | 25 | 35 | 1.024 |

The mechanical properties for weld and base metals are showed in Table 1.2, as per aforementioned standards [7, 9]. It was observed that the weld bead exhibited greater mechanical resistance than base metal, which complied with the basic design of welded joints.

The robotic GMAW process was performed with the three parameters [1, 2] listed in Table 1.3, which can be represented by the heat input that was calculated by Eqs. (1.2) and (1.3) [4] for each double T-welded joint under two different conditions: medium (CM13) and long (CL1) weld passes.

$$Q_{net} = Q_{arc} \tag{1.2}$$

where $Q_{net}$ = net heat input (kJ/mm), $\eta$ = heat transfer efficiency, 0.85 for GMAW process,

$$Q_{arc} = \left(\frac{EI}{s}\right)60 = \text{arc energy}\left(\text{kJ / mm}\right), \tag{1.3}$$

$E$ = arc voltage (V), $I$ = welding current (A), and $s$ = travel speed (cm/min).

Specimens of welded joints were machined according to the Vickers hardness and Charpy impact tests and microscopy. For the evaluation of micro-hardness, transversal specimens of 10 mm in width and 40 mm in length were used. In order to characterize the impact energy of the weld and base metals, test pieces were machined with reduced type-A, rectangular cross section, transverse orientation, 2.5 mm in thickness, 10 mm in width, 55 mm in length, and V-notch at 45°.

Vickers hardness testing was performed with an Emco Test Durascan micro-hardness tester using Ecos Workflow software applying a 4.9 N load and repeating every reading ten times for each microstructural zone as per ASTM E384 [11]. A Charpy test was carried out with a Tinius Olsen pendulum impact machine at 20 °C due to the solar service conditions and in triplicate for weld and base metals, in accordance with ASTM E23 [12]. For microstructural characterization, the specimens were cut in pieces 40 mm in length and 10 mm in width, ground, polished with alumina, and subsequently etched using Nital 2 % to reveal the three microstructural zones studied. The microstructural features were observed using a CARL ZEISS-Gemini Supra 55 VP SEM operated at 20 kV. The fractography of impact fractured specimens of weld and base metals was performed by the same SEM operated at 10 kV.

## 1.3 Results and Discussion

The robotic welded joints of S235JR carbon steel obtained at a greater heat input generated by a lower welding speed exhibited larger HAZ width, higher volume fraction in the weld bead, as well as a larger hardening of the weld bead compared to the HAZ. The hardening in the weld bead produced a reduction of impact energy and small microvoids distributed throughout the impact fracture surfaces showing ductile failure as the base metal.

### *1.3.1 Stereoscope Macrostructural Analysis*

Stereoscope observations performed on the longitudinal welded joints with medium and long weld passes revealed three well-defined microstructural zones as is shown in Fig. 1.1. These zones were identified as base metal outside the joint, HAZ as a band between weld bead and steel plates within the fusion line with a width of 1.65 mm for a medium weld pass (Fig. 1.1a) and 1.84 mm for long pass (Fig. 1.1b). Weld metal with beading at each side of the T-joint shows columnar macro-grains and major welding penetration for long passes due to higher heat input [1].

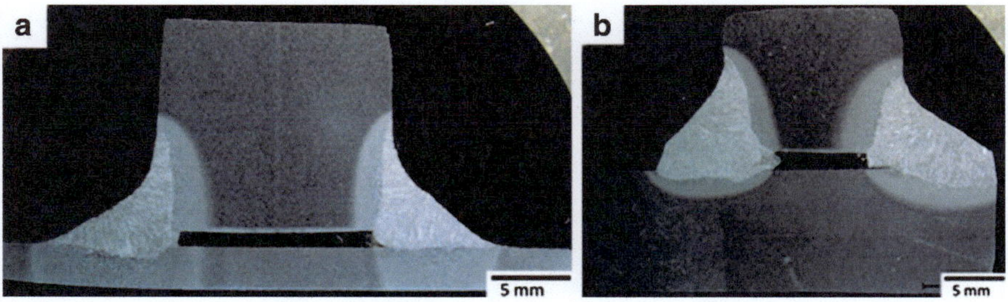


**Fig. 1.1** Macrostructures of GMAW T-joints showing three microstructures and width of HAZ: (**a**) medium weld passes and (**b**) long passes

### 1.3.2 SEM Microstructural Analysis

According to SEM observations on the specimens with long and medium weld passes, there were similar microstructures with differences in amounts and sizes of microstructural phases in the weld metal and HAZ. In the welded joint containing medium weld passes, the weld bead exhibited a heterogeneous solidification microstructure formed by columnar ferritic grains (dark zones) in the direction of the heat flow [4], which showed Widmastätten ferrite sideplates [13] growing from columnar grains. Among the long grains was observed fine acicular ferrite in the form of random needles (Fig. 1.2a). The most critical zone is the HAZ generated by the recrystallization process, resulting in two recrystallized zones: first $HAZ_1$ near the base metal and the second $HAZ_2$ next to the fusion line, which was composed of a considerable amount of coarse-grained ferrite (white areas) [6] and minimum amount of acicular ferrite, Fig. 1.2b. The $HAZ_1$ showed a recrystallized ferrite matrix with medium sized grain and evidence of thermally affected pearlite (white zones), Fig. 1.2c.

For the specimen with long weld passes, it was found that the weld metal exhibited larger amounts of acicular ferrite needles (white areas) between fine enlarged grains which exhibited higher amounts of fine Widmastätten ferrite sideplates [13] (Fig. 1.3a). At the $HAZ_2$ there was fine-grained ferrite (dark areas) [4] as well as an increase in the amount of fine acicular ferrite (white zones), Fig. 1.3b. The $HAZ_1$

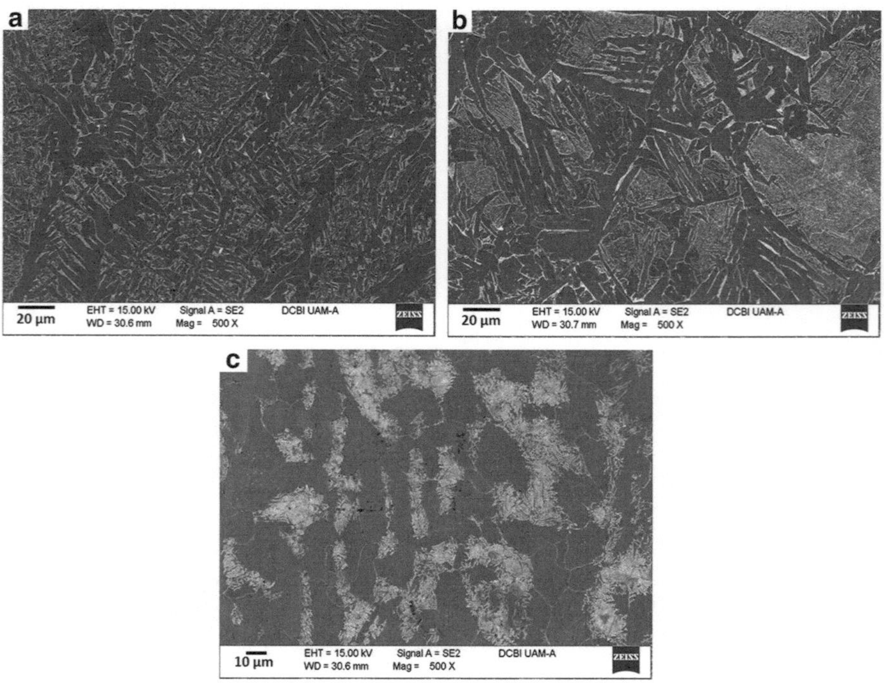


**Fig. 1.2** SEM micrographs of T-welded joint with medium weld passes: (**a**) columnar ferrite grains in weld bead, (**b**) coarse-grained ferrite in $HAZ_2$, and (**c**) heat affected pearlite in $HAZ_1$

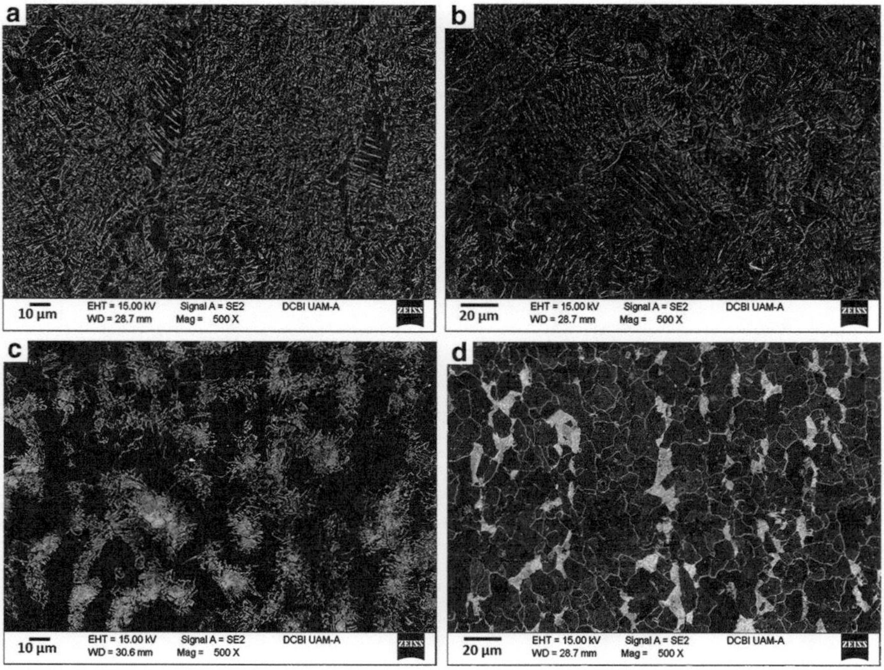


**Fig. 1.3** SEM micrographs of T welded joint with long weld passes: (**a**) fine Widmastätten ferrite in weld bead, (**b**) fine-grained ferrite in $HAZ_2$, (**c**) large amount of heat affected pearlite in $HAZ_1$ and (**d**) unaffected pearlite in base metal

exhibited a recrystallized ferrite matrix and a clearly larger amount of pearlite with a higher degree of thermal affectation (white areas), as can be seen in Fig. 1.3c compared to unaffected pearlite bands in the base metal with laminated microstructure at the hot rolling direction (Fig. 1.3d).

These microstructural features observed in both weld conditions would affect the mechanical resistance of the welded joint, hardness as well as fracture behavior under impact load.

### *1.3.3 Volume Fraction of Microstructural Regions*

The quantity of complex microstructural ferritic phases observed in the weld bead and complete HAZ was measured as a volume fraction [4], which is reported in Fig. 1.4a. We can see that the higher heat input favored greater volume fraction in both microstructural zones generated after the robotic welding cycle. The weld metal reached a larger volume fraction with greater heat input compared to the HAZ, which was associated with a lower welding speed resulting in a larger amount of Widmastätten ferrite. However, the critical HAZ exhibited a greater increase in

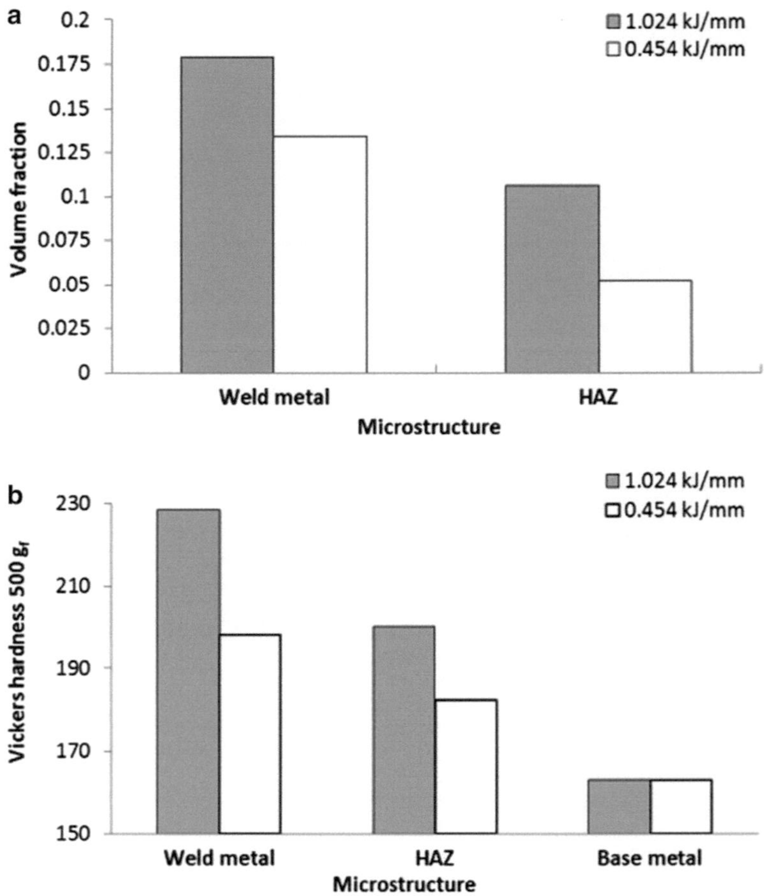


**Fig. 1.4** (**a**) Volume fraction of ferritic phases and (**b**) micro-hardness of weld metal and HAZ as a function of heat input

volume fraction (104 %) than the weld metal (34 %) as a consequence of greater heat input [4] generating an increase in width (12 %) and a greater susceptibility to cold cracking [5, 6]. This microstructural behavior was indirectly related to the change in micro-hardness explained in the following section.

### *1.3.4 Vickers Hardness Behavior*

The variation in micro-hardness with heat input for the weld bead and HAZ is showed in Fig. 1.4b. It can be seen that the weld bead reached higher Vickers hardness for both speed conditions compared to the HAZ and the base metal [4]. Greater heat input produced greater hardness for the weld metal [14] followed by the HAZ and ductile base metal showing less hardness for the T-welded joint with long weld passes in comparison with welded joint having medium passes. Moreover, the weld metal reached a major increase in hardness (15 %), i.e., moderate hardening compared to the HAZ with increment of 10 %, which was linked to the heterogeneous microstructure formed by columnar grains containing Widmastätten ferrite sideplates [13] and separated by acicular ferrite. This hardening was favored by the increase in current (26 %) and reduction of welding speed (60 %), resulting in a greater heat input with an increase of 226 %. The Vickers hardness for base metal did not change with heat input due to it not being affected by the heat generated during the robotic welding thermal cycle.

Therefore, the hardness and volume fraction must be reduced in both microstructural zones to assure their resistance to cold cracking in service conditions using a welding speed higher than 35 cm/min. This hardening behavior was indirectly related to the fracture changes.

### *1.3.5 Fractography*

The impact results for the weld bead indicated a significant 26 % reduction in the impact fracture energy (117 J) with a greater heat input compared to weld obtained with a less heat input (160 J) and base metal (193 J). This energy behavior produced changes in the surface fractures. Additional indirect measurements of the changes involving the increase in volume fraction and hardening were achieved by SEM study of surfaces obtained from the impact fractured specimens for the base and weld metals obtained with different heat input. One of the interesting results for all fractures was that the surface exhibited microvoids morphology characteristic of ductile type failure associated with the mechanism of nucleation, growth and coalescence of microcavities, as can be seen in Fig. 1.5. The cavity nucleation centers were promoted by the presence of non-metallic inclusions with different sizes, shapes, and distributions. Circular inclusions were observed for weld beads (Fig. 1.5a, b) and elongated inclusions in base metal (Fig. 1.5c).

The effect of the increase in heat input on the fracture surfaces for both conditions of weld provoked a significant change in the size and amount of microvoids. As such, an appreciable reduction in the size of shallow microvoids and an increase in their number across the surface were observed on the fractures with greater heat input (Fig. 1.5a) compared to surfaces for lesser heat input (Fig. 1.5b). A notable difference in larger size and lower density of microvoids was observed on the fractures of base metal (see Fig. 1.5c), when compared to specimens of both kinds of weld. This fracture behavior was related to the hardening produced by a higher heat input resulting in small microvoids on the fractures for the weld bead compared to surfaces of weldings produced by lower heat input with less hardness containing medium microvoids. Besides, the fractures of base metal showed larger microvoids which were linked to ductile behavior.

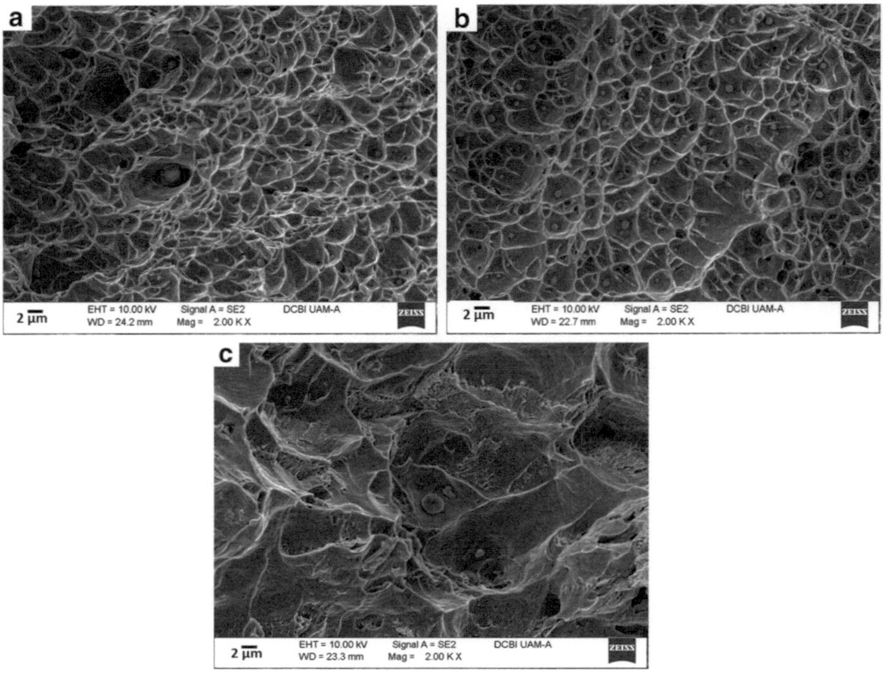


**Fig. 1.5** Fractography of impact fractured specimens showing microvoids in different sizes and amounts for: (**a**) weld bead with higher heat input, (**b**) weld with lower heat, and (**c**) base metal

## 1.4 Conclusions

T-welded joints obtained by robotic metal inert gas process with higher net heat input generated four notable changes including hardening, variation of microstructural volume fraction, reduction in Charpy energy, and smaller size of microvoids on the ductile fracture surface. The welded joint containing the long weld bead was obtained with greater heat input due to lower welding speed resulting in the hardening as a consequence of greater volume fraction. The weld bead reached greater hardening than the HAZ associated with the presence of complex microstructural phases such as the fine acicular ferrite between narrow columnar grains containing Widmastätten ferrite. The recrystallized HAZ was the most affected microstructure due to major increase in volume fraction and width compared to weld bead linked to the heterogeneous fine-grained ferrite and fine acicular ferrite giving greater susceptibility for cracking affecting the joint weldability. The main scientific and technological contribution of this work is showing that the fractography of impact fracture surfaces of weld bead exhibited minor impact energy and evident ductile failure containing smaller microvoids in larger quantities compared to weldings produced with lower heat input and base metal.

## References

1. Kumar BP (2012) Effect of welding parameters of gas metal arc welding on weld bead geometry: a review. Int J Eng Sci Technol 4(7):3446–3449
2. Lee D, Ku N, Kim WT, Kim J, Lee YK, Son SY (2011) Development and application of an intelligent welding robot system for shipbuilding. Robot Comput Integr Manuf 27(2): 377–388
3. Karadeniz E, Ozsarac U, Yildiz C (2007) The effect process parameters on penetration in gas metal arc welding processes. Mater Des 28(2):649–656
4. González-Gutiérrez S, Vargas-Arista B, Solís J, García-Vázquez F (2010) Effect of wire feed rate on the microstructure and microhardness of multilayer weldment by GMAW process on ASTM A633 steel. 32 Congreso Internacional de Metalurgia y Materiales, Coahuila, México, pp 1–8
5. DIN-EN 1011-2 (2001) Recommendations for welding of metallic materials, part 2: arc welding of ferritic steels. DeutschesInstitut fur Normung, Germany, p 9
6. UNEEN ISO 15609-1 (2004) Specification and qualification of welding procedures for metallic materials. Part 1: arc welding. AENOR, Spain, pp 6–10
7. AWS A5.18 (2005) Specification for carbon steel electrodes and rods for gas shielded arc welding. American Welding Society, USA, pp 2–6
8. DIN EN 10025-2 (2004) Hot rolled products of structural steels, part 2: technical delivery conditions for non-alloy structural steels, Germany, pp 18–29
9. DIN EN 10025-1 (2004) Hot rolled products of structural steels, part 1: general technical delivery conditions, Germany, p 10
10. Salazar Garrido JA (2009) Análisis microestructural mediante ultrasonido en una unión de soldadura y su evaluación de propiedades en fatiga. Tesis de Maestría en Tecnología de Soldadura Industrial. COMIMSA, México, pp 32–34
11. ASTM E384 (2005) Standard test method for microindentation hardness of materials. ASTM International, USA, pp 1–8
12. ASTM E-23 (2006) Standard test methods for notched bar impact testing of metallic materials. ASTM International, USA, pp 2–8
13. Thewlis G (2004) Classification and quantification of microstructure in steels. Mater Sci Technol 20(2):143–160
14. Ibrahim IA, Mohamat SA, Amir A, Ghalib A (2012) The effect of gas metal arc welding (GMAW) processes on different welding parameters. Eng Procedia 41:1502–1506

# Chapter 2
# Distortion Evaluation in Dissimilar Stainless Steel Joints Welded by GMAW Process

**A.F. Miranda Pérez, G.Y. Pérez Medina, E. Hurtado Delgado, and F.A. Reyes Valdés**

**Abstract** Manufacturing in almost all industries involves joining together many components. Welding is the most popular joining process, which is cost effective, reliable, and versatile. In the automobile industry, when welding is applied, distortion can occur and produce additional costs of rework. Distortion became a potential problem in this process, since the joint undergoes significant thermal expansion and contraction, longitudinal and transverse shrinkages and leads to a non-desired welded structure. Recently, the usage of more lighted alloys in automobile sector ensures the conductor's safety and reduction of emission; stainless steels have been placed in this industry as a compelling alternative for some carbon steels.

The employment of hybrid structures and components is willing field in vehicles. In this study dissimilar joints of austenitic and ferritic stainless steels welded by robotic gas metal arc welding process were microstructurally examined by optical and scanning electron microscope. Austenite twins are presented in the heat affected zone (HAZ) on the austenite plate, on the other hand allotriomorphic ferrite was highlighted in the transition zone. Microhardness test was performed along the cross-section of the weld. Thermocouples were employed during welding in order to determine the temperature distribution and analyze the thermal cycles from the weld to the base metal. Distortion evaluations were carried out by employing GeoMagic® Design™ X and Konica Minolta Range 7 for weld scanning and the deviation of the plates was calculated.

**Keywords** Stainless steels • Distortion • Microstructural characterization • GMAW • Welding

---

A.F. Miranda Pérez (✉) • G.Y. Pérez Medina • E. Hurtado Delgado • F.A. Reyes Valdés
Corporación Mexicana de Investigación en Materiales, Ciencia y Tecnología 790, Saltillo, Coahuila, México
e-mail: argelia.miranda@comimsa.com; eduardohd@comimsa.com

 
R. Pérez Campos et al. (eds.), *Materials Characterization*,
DOI 10.1007/978-3-319-15204-2_2

## 2.1 Introduction

The automotive sector has been regarded to develop safe vehicles, reduce weight in their components, and increase the fuel efficiency. In order to achieve these characteristics, the materials selection and manufacturing of them is an essential issue. Welding is seen as one of the most profitables of production to joint metals. During welding, the heating and cooling rates in the base metal and seam weld are not equal, and this can cause some distortion to the entire assembly [1, 2]. This mechanism can affect the dimension accuracy and the structure behavior, which can cause costly repair work or even replicate of the work [3].

According to Folkhard [4], three groups of factors exist that can cause distortion during welding, the materials factors including the physical and mechanical properties, the design factors including type of the joint, dimensions and thickness, and the third one which covers the welding method.

As can be seen, there are too many factors that may cause distortion. The usage of stainless steels (SS) in the vehicle industry is increasing; if deformation can be predicted in a welded structure to reduce or eliminate distortion, these predictions can be advantageous for line production [5]. Ferritic stainless steels are particularly employed in automotive exhaust systems due to their ability to improve high temperature properties [6, 7]. However, their structure can be rather brittle which means poor weldability, introducing austenitic stainless steels with marked advantages with regard to toughness and good weldability [8].

One joining process more employed in automotive sector is gas metal arc welding (GMAW), using either short circuiting, pulsed-arc or spray-arc transfer. The process requires the use of a filler metal as a metallic wire with shielding gas as protection from corrosion of the molten metal [9]. Shanmugam et al. [10] studied dissimilar joints between ferritic stainless steels and low carbon steel obtaining in the microstructure martensite laths and few ferrite, upper bainite and widmanstatten ferrite in the heat affected zone (HAZ).

The aim of the research was to evaluate the distortion of dissimilar stainless steels of different geometries, in order to analyze the behavior of the microstructure and the effect of the temperature distribution of the welded plates. In this study a dissimilar joint of two different types of stainless steels 410-S ferritic SS and 304-L austenitic SS using robotic GTAW as welding process were characterized using optical and scanning electron microscopes (SEMs).

## 2.2 Experimental

Ferritic and austenitic stainless steel plates of 5.7 and 4.6 mm of thickness, respectively, were employed in this study (Fig. 2.1). Their chemical composition, mechanical and thermal properties are shown in Tables 2.1 and 2.2.

Robotic GMAW process short circuiting was employed to join the experimental material using KR16 KUKA model in the manufacturing cell at COMIMSA. The welding conditions are shown in Table 2.3.

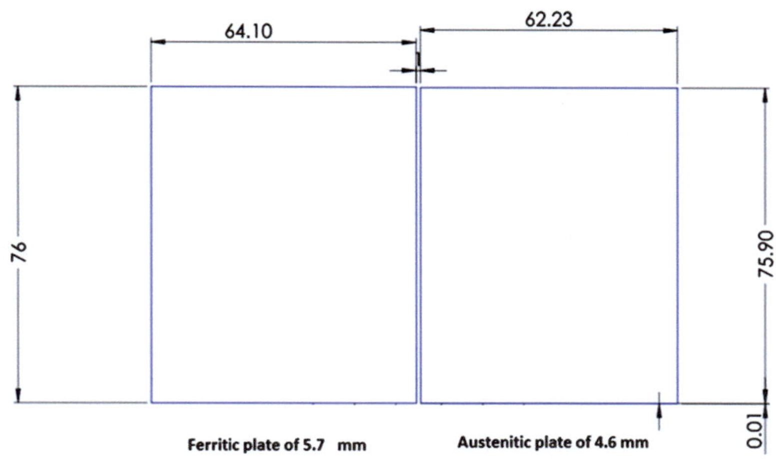


**Fig. 2.1** Schematic illustration of the stainless steel plates

**Table 2.1** Chemical composition wt.% of ferritic and austenitic stainless steels

| Base metal | C | S | Mn | P | Si | Cr | Ni | Mo | Cu | V |
|---|---|---|---|---|---|---|---|---|---|---|
| Ferritic 410-S | 0.01 | 0.01 | 0.47 | 0.03 | 0.04 | 13.1 | 0.17 | 0.21 | 0.07 | 0.03 |
| Austenitic 304-L | 0.04 | 0.01 | 1.55 | 0.03 | 0.46 | 18.26 | 8.85 | 0.31 | 0.41 | 0.09 |

**Table 2.2** Mechanical and thermal properties of ferritic and austenitic stainless steels

| Type | Yield strength (MPa) | Elongation 2″ (mm) | Hardness (Hv) | Heat capacity (J/g °C) | Thermal conductivity [W/m-K (500°)] |
|---|---|---|---|---|---|
| 410-S | 200 | 33 | 130 | 0.46 | 28.7 |
| 304-L | 170 | 40 | 185 | 0.5 | 21.5 |

**Table 2.3** GMAW parameters

| Electrode | ER-308 L (AWS) |
|---|---|
| Voltage | 22 V |
| Gas (Ar) | 100 % |
| Weld speed range | 0.9 m/s |

Schematic illustration of the thermocouples position is presented in Fig. 2.2; two different acquisition cards were employed in order to obtain more accurate results of the temperature distribution along the weld.

Scanner Konica Minolta Range 7 was employed since the capture data of any industrial part can be displayed on a computer and compared with 3D CAD models. Once the CAD model was acquired, GeoMagic® Design™ X was used to measure the distortion of the butt weld.

All metallographic samples were prepared using standard techniques of mechanical polishing; the specimens were then etched with aqua regia solution. In order to determine the chemical composition of the phases observed in the microstructure, an energy dispersive X-ray spectrometer (EDX) system fixed to the SEM was employed.

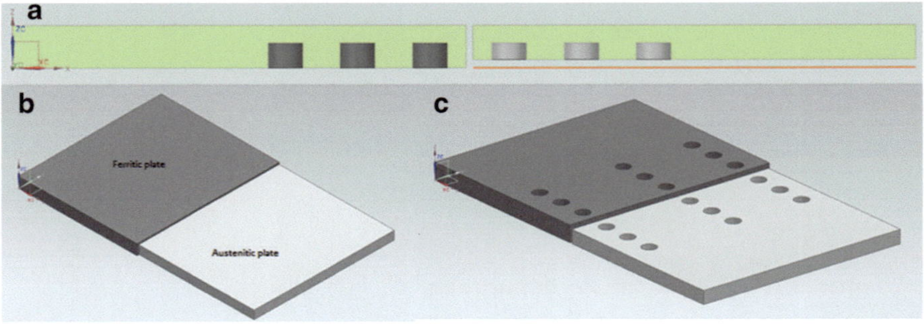


**Fig. 2.2** Illustration of the stainless steel plates: (**a**) *front view* of the butt joint plates, (**b**) ferritic and austenitic stainless steel plates, (**c**) blast-holes for the thermocouples lectures

## 2.3 Results and Discussion

### *2.3.1 Base Metal*

The austenitic stainless steel microstructure is presented in Fig. 2.3a, the optical micrograph is constituted by austenite matrix within austenitic twins. Moreover, Fig. 2.3b evidences the ferritic 410S stainless steel with ferritic matrix, whose ferritic grains are elongated along the longitudinal section.

### *2.3.2 Dissimilar Welding*

Figure 2.4 evidences the microstructure of the dissimilar joint between the ferritic and austenitic stainless steels, as can be appreciated on the austenitic side the HAZ results smaller than the ferritic side. However, full penetration for both sides was obtained. From the flat plates presented in the experimental part, it could be seen some distortion at the austenitic side after welding.

Figure 2.5 shows from the microstructure of the plates in the heat affected-transition-fusion zone. In the austenitic plate, starting from the fusion boundary, dendritic grains along the weld are presented in Fig. 2.5a. In the second micrograph (Fig. 2.5b) the transition zone is highlighted showing the HAZ with coarse austenitic grains and in Fig. 2.5c HAZ at higher magnification evidence the austenitic matrix within austenitic twin grains with higher carbon content.

Figure 2.6 evidences the ferritic stainless steel side of the joint. In Fig. 2.6a HAZ is shown with coarse grains in a ferritic matrix [11], some whiter grains are presented in the microstructure which is explained using the EDS results. On the transition zone some allotriomorphic ferrite is formed from the fusion zone to the HAZ.

In addition, the distortion analysis was performed using GeoMagic® Design™ X program which is presented in Fig. 2.7. The contraction on the austenitic plate for almost 5 mm according to the program results can be appreciated.

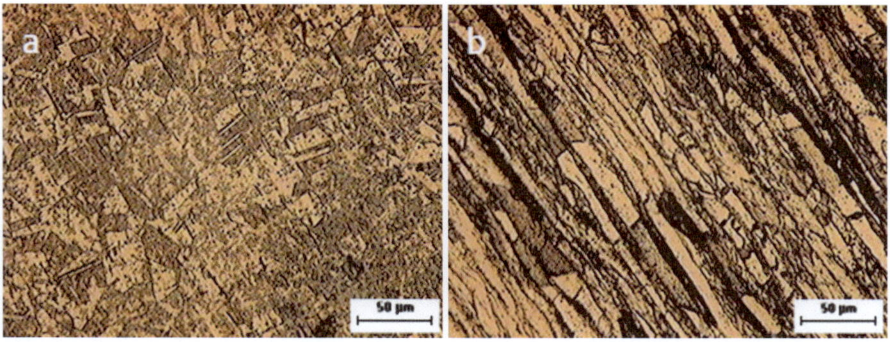


**Fig. 2.3** Optical micrograph: (**a**) austenite matrix base metal, (**b**) ferritic matrix base metal

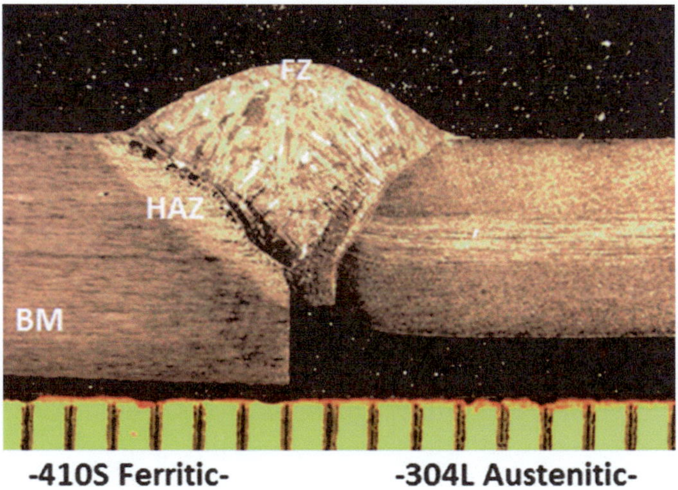


**Fig. 2.4** Stereoscope micrograph of the dissimilar joint

Thermocouples were used in order to observe the temperature distribution; the channel lectures are displayed in a plot form in Fig. 2.8.

### 2.3.3 *EDS Analysis*

The selected points for chemical analysis at the current phases from both steels are presented in Fig. 2.9 and the resultant chemical composition is evidenced in Table 2.4.

On the HAZ for the ferritic plate was noticed some white grains mentioned before, which are rich in carbon; this could be due to the seam weld employed during welding and the carbon diffusion [12, 13]. Besides, according to Table 2.4 chromium element decreases.

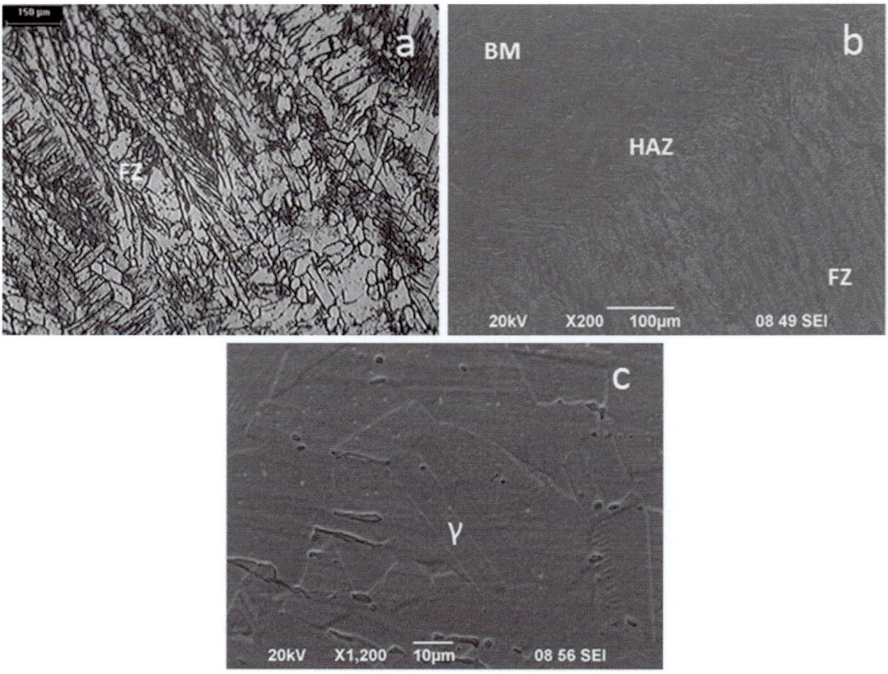


**Fig. 2.5** (**a**) Image obtained by optical microscopy of fusion zone; (**b**) SEM micrograph of the transition zone on the austenitic side base metal, heat affected zone and fusion zone; (**c**) SEM micrograph austenitic HAZ

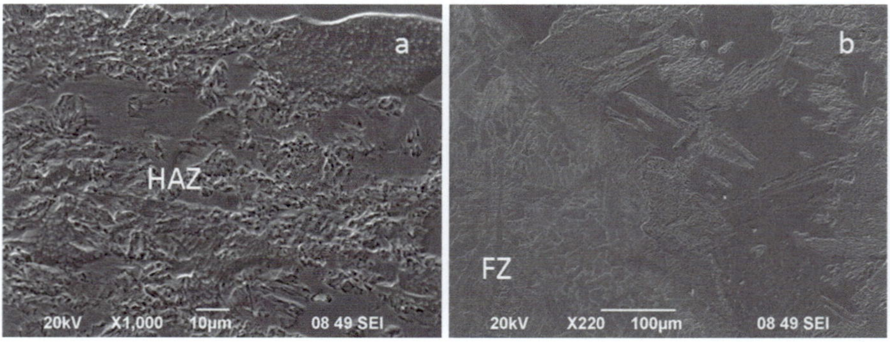


**Fig. 2.6** SEM micrograph of the ferritic side of the weld: (**a**) HAZ and (**b**) FZ

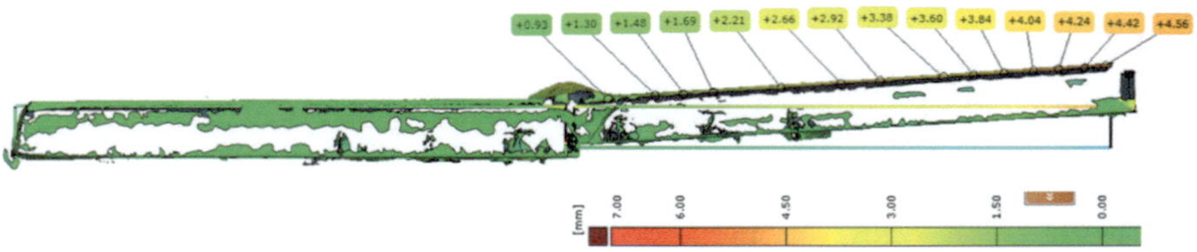


**Fig. 2.7** Schematic view of the distortion evaluated at different points

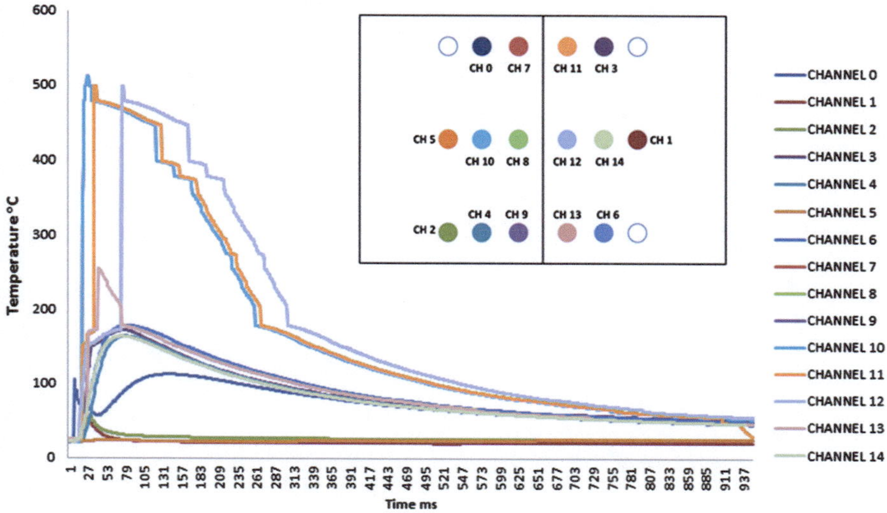


**Fig. 2.8** Thermocouples measurements of the different channels located in the weld

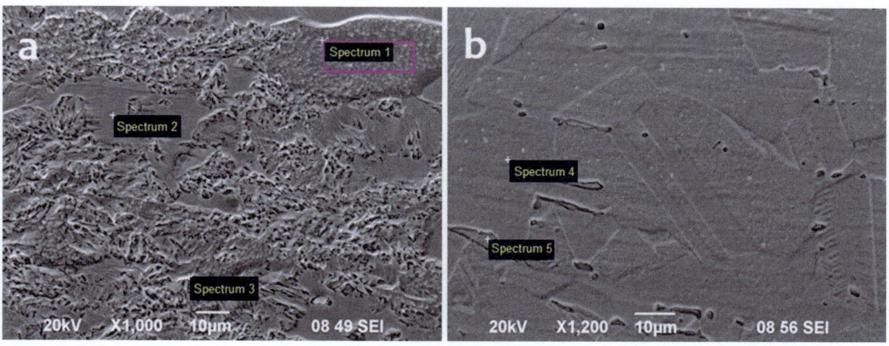


**Fig. 2.9** SEM micrograph with the spectrum for the EDS analysis: (**a**) ferritic stainless steel and (**b**) austenitic stainless steel

**Table 2.4** EDS analysis with the chemical composition (wt.%)

| | Cr | Fe | Ni | Mo | Mn |
|---|---|---|---|---|---|
| Spec1 | 10.03 | 72.01 | 0.38 | 0.39 | |
| Spec2 | 9.86 | 69.45 | 0.38 | 0.02 | |
| Spec3 | 10.72 | 69.13 | 0.46 | | |
| Spec4 | 17.84 | 56.39 | 4.18 | | 1.36 |
| Spec5 | 13.74 | 53.92 | 6.89 | | 1.28 |

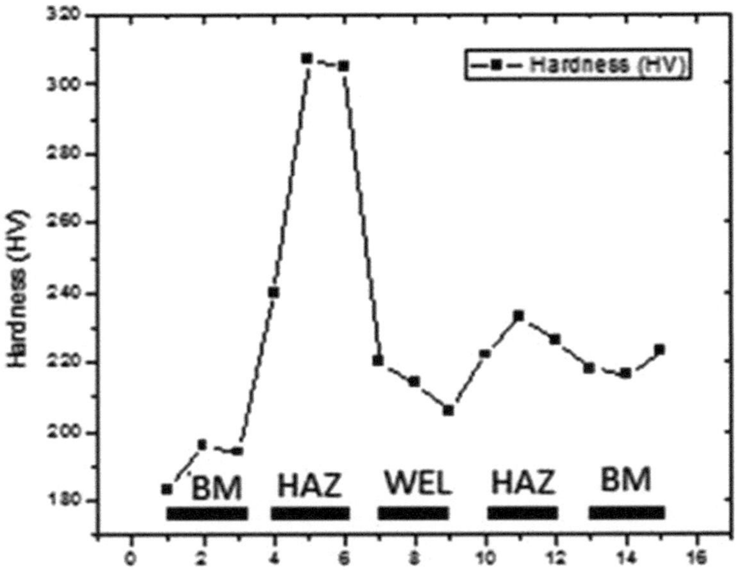


**Fig. 2.10** Microhardness profile of dissimilar joint

### *2.3.4 Microhardness Analysis*

Microhardness profiles of the welded regions indicate that the hardness in HAZ of ferritic stainless steel was relatively elevated, nearly 350 Hv compared with 230 Hv in the HAZ of austenitic steel (Fig. 2.10), indicating the development of coarse grains in a ferritic matrix and white grains rich in chromium in accordance with the EDS analysis. On the other hand, the behavior of microhardness in the HAZ of austenitic steel was related to the filler metal without significant changes. Nevertheless, the distortion was present in austenitic thin layer.

## 2.4 Conclusions

For the ferritic 410-S and austenitic 304-L SS dissimilar welded plates using GTAW robotic process was carried out. The microstructural characterization concludes as follows:

- The dissimilar joint presented on the austenitic side, entire austenitic matrix on base metal, even though some coarse austenitic grains at HAZ were highlighted, and austenite dendrites in the FZ were observed. Instead, for the ferritic zone carbon increment in the chemical composition was evidenced giving white grains rich in Cr and C.

- Distortion evaluation reached almost 5 mm of contraction in the austenitic plate according to GeoMagic® Design™ X program. This is due to thickness of the ferritic stainless steel plate, increasing the amount of weld, therefore the solidification shrinkage and the thermal contraction increase as well.
- The software employed during the experiment resulted beneficial for distortion measurements. However, scanning meshing must be conducted well in order to obtain accurate results.
- Microhardness evaluation evidenced the increment in the HAZ on the ferritic plate indicating the development of coarse grains in a ferritic matrix and white grains rich in chromium.

**Acknowledgments** The authors would like to thank J. C. Montoya Betancourt and L. A. Lara Arredondo for their technical support, which was most helpful in conducting some of the experiments.

## References

1. Lee C-H, Chang K-H (2012) Temperature fields and residual stress distributions in dissimilar steel. Appl Therm Eng 45–46:33–41
2. Deng D, Murakawa H (2008) FEM prediction of buckling distortion induced by welding in thin plate panel structures. Comput Mater Sci 43(4):591–607
3. Deng D, Zhou Y, Bi T, Liu X (2013) Experimental and numerical investigations of welding distortion induced by $CO_2$ gas arc welding in thin-plate bead-on joints. Mater Des 52:720–729
4. Folkhard E (1988) Welding metallurgy of austenitic stainless steels. In: Welding metallurgy of stainless steels. Springer, Vienna, pp 197–222
5. Hänninen H, Aaltonen P, Brederholm A, EhrnstÈn U, Gripenberg H, Toivonen A, Pitkänen J, Virkkunen I (2006) Dissimilar metal weld joints and their performance in nuclear power plant and oil refinery conditions. VTT research notes 2347
6. Kim JK, Hong SG, Kang KB, Kang CY (2009) Microstructure and high temperature properties of the dissimilar weld between ferritic stainless steel and carbon steel. Met Mater Int 15(5):843–849
7. Srinivasan K, Balasubramanian V (2011) Effect of heat input on fume generation and joint properties of gas metal arc welded austenitic stainless steel. J Iron Steel Res Int 18(10):72–79
8. Tian L, Luo Y, Wang Y, Wu X (2014) Prediction of transverse and angular distortions of gas tungsten arc bead-on plate welding using artificial neural network. Mater Des 54:458–472
9. Samal MK, Seidenfuss M, Roos E, Balani K (2011) Investigation of failure behavior of ferritic–austenitic type of dissimilar steel welded joints. Eng Fail Anal 18:999–1008
10. Shanmugam K, Lakshminarayanan AK, Balasubramanian V (2009) Effect of weld metal properties on fatigue crack growth behaviour of gas tungsten arc welded AISI 409M grade ferritic stainless steel joints. Int J Press Vessel Pip 86:517–524
11. Ha TK, Jeong HT, Sung HJ (2007) High temperature bending fatigue behavior of stainless steels for automotive exhaust. J Mater Process 555:187–188
12. Wu W, Hu S, Shen J (2014) Microstructure, mechanical properties and corrosion behavior of laser welded dissimilar joints between ferritic stainless steel and carbon steel. Mater Des 65:855–861
13. Manurung YHP, Lidam RN, Ridzwan Rahim M, Yusof Zakaria M, Ridhwan Redza M, Shahar Sulaiman M, Tham G, Abas SK (2013) Welding distortion analysis of multipass joint combination with different sequences using 3D FEM and experiment. Int J Press Vessel Pip 111–112:89–98

# Chapter 3
# 3-D Porosity in T-Welded Connections Repaired by Grinding and Wet Welding

**G. Terán, S. Capula-Colindres, R. Cuamatzi-Meléndez, D. Angeles-Herrera, and A. Albiter**

**Abstract** This work presents a methodology to create 3-D porosity into finite element models of T-welded connections repaired by rectangular grinding and filled with wet welding. In a wet welding process, the porosity increases with water depth, as a result, the developed wet welding can present low mechanical properties. T-welded connections are employed to fabricate fixed platform, which are used to exploit hydrocarbons located in shallow waters and because they are subjected to dynamic loads, so cracking can occur in such connections. Therefore, a combination of grinding and wet welding repair techniques can be used to repair damage in such offshore structures. To define the level of porosity, it was performed a metallographic analysis at the weld toe of T-welded connections which were grinded with a width of 4 mm and two grinding depths of 6 and 10 mm using conventional grinding equipment. Then, those grinded regions were filled by wet welding employing the shielded metal arc welding (SMAW) and a hyperbaric chamber simulating the different water depths: 50, 70, and 100 m. To apply the wet weld beads, there were employed E6013 coated electrodes with vinyl varnish. So, for the porosity analysis, samples were extracted from T-welded connections, which were analyzed employing an optical microscopy (OM) and scanning electronic microscopy (SEM). Results showed that the porosity can appear in spherical and elliptical shape in low part of every weld beads, and the pores are aligned at the bottom of the lower weld beads, while in the upper weld bead there was no porosity. The experimentally analyzed levels of porosity were created in the 3-D finite element models of the T-welded connections in the wet weld beads.

**Keywords** 3-D porosity • T-welded connections • Grinding • Wet welding • Platforms

G. Terán (✉) • S. Capula-Colindres • R. Cuamatzi-Meléndez • D. Angeles-Herrera • A. Albiter
Instituto Mexicano del Petróleo, Eje central Lázaro Cárdenas 152, Col. San Bartolo Atepehuacan, México D.F. 07730, México
e-mail: gerardoteranm@gmail.com; selenecapula@gmail.com; rcuamatzi@imp.mx; dangelesh600@alumno.ipn.mx; aalbiter@imp.mx



R. Pérez Campos et al. (eds.), *Materials Characterization*,
DOI 10.1007/978-3-319-15204-2_3

## 3.1 Introduction

T-welded connections are employed and used to build fixed platforms in the Gulf of México. The platforms are subjected to dynamic loads, so these platforms need maintenance to complete and/or extend their service life in order to continue producing oil and gas in shallow waters [1]. Therefore, a combination of grinding and wet welding has been used to repair localized damage at the weld toe of the T-welded connections due to its simplicity compared to the welding habitat [2]. Although there are various studies to improve mechanical properties of the weld beads in a wet welding process, there are still various problems to deal with. The main problems are the porosity and cracking that occurs in the weld pool, which is reflected in low mechanical properties of wet the weld beads. An example is the hydrogen produced from the dissociation of water leading to the formation of defects, such as pores and cracks in the weld bead (WB) and in the heat affected zone (HAZ). Such problems can lead to low ductility in the WB [3]. These two problems are the main causes for the mechanical properties of the wet welding can not be comparable with laboratory conditions.

The porosity formation in wet welding has been studied to understand its mechanism formations [4, 5], and there are works to understand the porosity process in wet welding. Padilla et al. [3] employed the X-ray microtomography method (μCT) and the image analysis (IA) image and measure 3D cracks in wet welding. They used A36 steel plates and 0.5 m water depth to obtain specimens with dimensions of $4 \times 4 \times 10$ mm. Parciornick et al. [6] performed work to characterize pores and cracks in wet welding by (μCT) and digital optical microscopy. They used commercial electrodes and A36 steel for 20 m water depth for two samples 5S1 and 1N1 for 20 and 0.5 m water depths, respectively. It was observed a pore alignment with major axis to the welding surface and an inclination due to the welding direction. Also, it is reported a pore fraction of 3.4 % for 3D images, and 2.2 % of porosity using traditional technique (metallographic). However, the main limitation of the μCT method is the power of the tomography, which is necessary to traverse large samples of a dense material, such as steel and the compromise between sample size and resolution. With this technique, it can only use samples of 2 mm of thickness. However, the lengths of wet beads are approximately 25–30 cm for E6013 coated electrode.

In the present work, it was performed experimental work to characterize the porosity using optical microscopy (OM) and scanning electron microscopy (SEM) in the T-welded connections repaired by grinding and wet welding. The T-welded connections were fabricated with A36 steel plates. The wet welding was performed at three different water depths: 50, 70, and 100 m in a hyperbaric chamber. The wet welding was applied employing E6013 coated electrodes with vinyl varnish. To created 3D pores, finite element method (FEM) was employed.

## 3.2 Experimental Work

This experimental work was made according to Terán in previous work [1, 7] and the T-welded connections were prepared from ASTM A36 steel in dry conditions. A longitudinal grinding with rectangular grinding profile at the weld toe and a hyperbaric

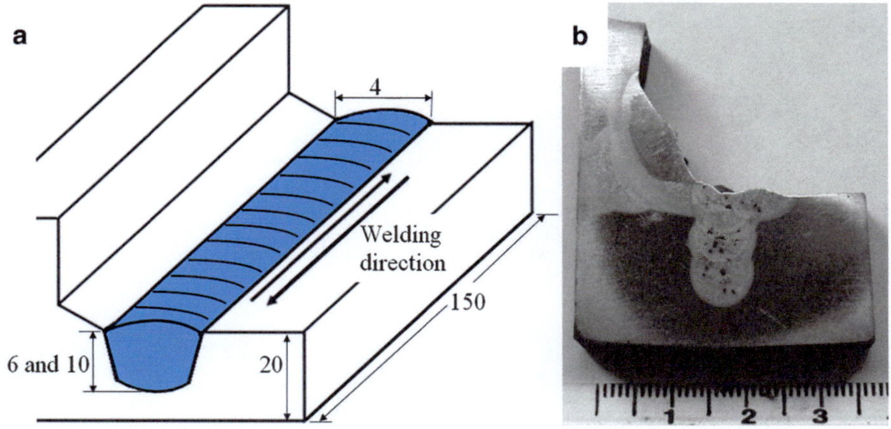


**Fig. 3.1** Schematic of T-welded connections, (**a**) dimensions (units in mm), and (**b**) sample extracted from T-welded connection, 10 mm grinding depth and 70 m water depth

chamber was employed to fill the grinded material by wet welding. In order to apply the wet welding, the gravity welding system was used inside the hyperbaric chamber. To employ wet welding, the shielded metal arc welding (SMAW) was used. Three different water depths at 50, 70, and 100 m were simulated. E6013 electrodes were coated with vinyl varnish and they were used. Samples for porosity analysis were extracted from T-welded connections. To obtain pores sizes and porosity level, a scanning electron microscope (SEM) Philips model XL 30 ESEM 25 KV and an optical microscope (OM) model from 5X to 50X were employed. The FEM was used to create 3D porosity models and the commercial software ANSYS version 12.1 [8] was chosen. It is because 3D models can be created with precision in dimensions. The dimensions of the T-welded connections (rectangular grinding profile and welding direction) and samples extracted to obtain pore sizes are shown in Fig. 3.1a, b, respectively.

## 3.3 Results and Discussion

The results of porosity analysis and pore size are presented in Table 3.1. In this table [1], the size and percent porosity increases. It is caused by the gases generated in the wet welding process [2]. On the one hand, Fig. 3.2 shows porosity distribution in the wet welding beads by SEM. It is observed that the porosity is greater for 10 mm grinding depth condition. Also, for this grinding depth, more weld beads were required to fill the rectangular grinding profile; 7 and 3 weld beads for 10 and 6 mm grinding depths, respectively. So, due to the above, the condition of 10 mm grinding depth and 70 m water depth was chosen to characterize 3D pores.

The values of A–A, B–B, and C–C in Table 3.1 were taken from the macrograph images. Figure 3.3a shows the measurement of these pores in macrograph. A line was traced from the top weld (last weld bead) to the bottom weld (first weld bead

**Table 3.1** Porosity analysis for the different working conditions

| Welding depth (m) | Grinding depth (mm) | Average porosity (%) | Average pores size (μm) | | Pores numbers, lines | | |
|---|---|---|---|---|---|---|---|
| | | | L | H | A–A | B–B | C–C |
| 50 | 10 | 3 | 356 | 255 | 2.6 | 3.6 | 2.1 |
| 70 | 10 | 4 | 459 | 264 | 3.0 | 6 | 3.0 |
| 100 | 10 | 10 | 506 | 476 | 3.8 | 6.6 | 5 |
| 50 | 6 | 2 | 437 | 370 | 1.5 | 2.5 | 2.1 |
| 70 | 6 | 4 | 453 | 453 | 2 | 2.8 | 2 |
| 100 | 6 | 8 | 517 | 398 | 3 | 3 | 3.5 |

*L* length, *H* height

**Fig. 3.2** Porosity distribution in the wet weld beads

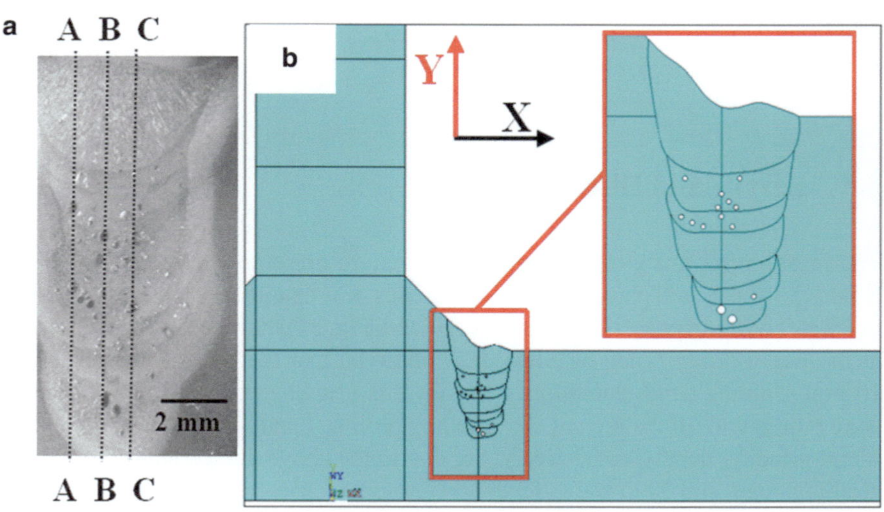


**Fig. 3.3** 2D porosity, (**a**) macrograph to measure pores and (**b**) 2D porosity with different wet welds beads, 10 mm grinding depth and 70 m water depth

deposited to fill the rectangular grinding). Then, pores were measured in the marked line. Four macrographs were obtained to measure the pores. It is noted that for the conditions of A–A, B–B, and C–C, the pores numbers are 3.0, 6.0, and 3.0, respectively. This means that for B–B line, there is a pore for every weld bead. In the top weld bead, porosity was not observed. For the next measurement in the macrographs, 3 pores for 6 weld beads were found. Therefore, with these measurements, 3D pores were created.

On the other hand, Table 3.1 shows that for the conditions of 70 m water depth and 10 mm grinding depth, the pore length and height are 459 and 265 μm, respectively. Then, to create 2D pores, it was considered a radius of 229.5 μm (459/2) for the larger pore, and 132 μm (264/2) to the smallest pore. It is noted that for the model in 2D, HAZ was not considered. 2D porosity is presented in Fig. 3.3b.

Once 2D porosity was performed, this was projected in 3D. This line (B–B) was performed in 3D. It is because the line B–B had a higher percentage of pores and it was at the center in the wet weld beads. Pores with radius of 132 μm were considered in the third plane (Z) in all wet welding beads. The total distance of the T-welded connections was 5.5 mm. Four macrographs were obtained to measure pores in the lines B–B. 6, 3, 6, and 3 pores were measured for the four lines B–B in macrographs. The total distance was divided into 20 parts, in order to insert 6 pores in line B–B, 3 pores in the next lines, and so on up to the 20 divisions. The division between each line was 549 μm. This separation was set so that the pore diameter (132 μm) could be performed with accuracy. Figure 3.4 shows a photograph for the 3D porosity distribution during wet welding beads.

In this way, Fig. 3.4 shows that in the two first weld beads (bottom), it was presented pore alignment. This pore alignment is caused due to the grinding depth (10 mm). It is half of the chord of the T-welded connection, and there is little space

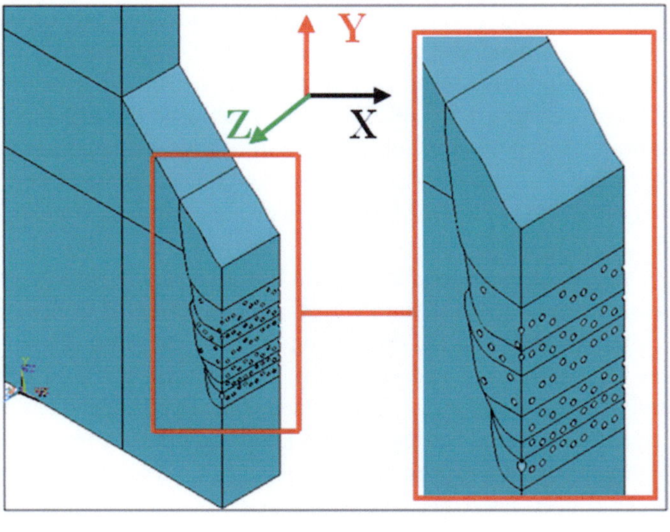


**Fig. 3.4** 3D porosity distribution generated in the wet welds beads for 70 m water depth and 10 mm grinding depth

for the gases ($H_2$, $CO_2$, and CO) to escape, and they are trapped in the wet weld beads. Also, gases have to break the hydrostatic pressure and the high speed rate cooling in the weld pool [1]. This generates greater porosity percentage. It is observed in the top weld beads that there is greater pore distribution. Also, there is more space between each pore in all wet weld beads, while in the top weld bead pores were not found. It is due in the top part of the weld beads there is enough space for porosity forming gases can escape. Also, for this 70 m water depth, the porosity percentage was 4 % and no pore was observed in the upper weld bead, see Fig. 3.3a. From the model of Fig. 3.4, it can be observed that a 3D visualization can help for a better understanding of the porosity formation and pore coalescence mechanism, also for a better defect quantification in wet weld beads.

In 3D works performed by Padilla et al. [3], it was found that the distance of the crack regions is in the range of 200–500 μm, and the classical methods to obtain porosity overestimate the true porosity. 3D visualization helps to estimate the mechanical properties and to quantify defects in the wet weld beads. The main limitation of the work reported by Padilla et al. [3] is the power of the tomography, which reaches resolutions of 2.17 μm for samples of 2 mm thickness. The wet welding beads have an average length of 25–30 cm for commercial electrodes E6013. It is needed a great number of slices for the porosity analysis.

Puchol et al. [9] presented a calculation methodology to obtain pores numbers in A36 steel for 50 and 100 m water depths. He found that there was a similarity in results between the theoretical and experimental values. Thus, with this work, pores numbers can be determined. Then, with the previous works, the pores percentage and pores numbers can be obtained to simulate water depths with E6013 coated electrodes and A36 steel.

The difference between models by the 3D finite elements models and the μCT method is that the FEM can make 3D models which can carry parameters, such as size and distributions of pores, etc. In this way, it is important to know the mechanical properties and determine what pore percentage the mechanical properties can be restored. On the other hand, the μCT technique only can be observed the pore distribution, density, alignment, size, etc. To make a porosity analysis, it is recommended as a first step to use the μCT method. Then used the FEM to include the mechanical properties, pore sizes, cluster area to restore the mechanical properties and compare them in laboratory conditions. As known, the ductility is reduced when critical pore size is reached [10]. So, with this problem, it cannot reach air conditions.

Finally, it can be seen the need to perform further research work to characterize full wet welding beads in 3D. Jean-Yves et al. [11] presented a methodology for generating refined meshes on the conversion of FEM data into 3D pores images, although this work is focused on cast Al alloy. Figure 3.5 shows a section of typical fracture surface for wet weld bead. It is presented pore alignment and cluster porosity. For a next research, it would be important to know the porosity fraction, the conditions in which the mechanical properties could be restored using different electrode types and water depths. These research works would help to establish safety criteria and apply to regulations in order to repair offshore structures.

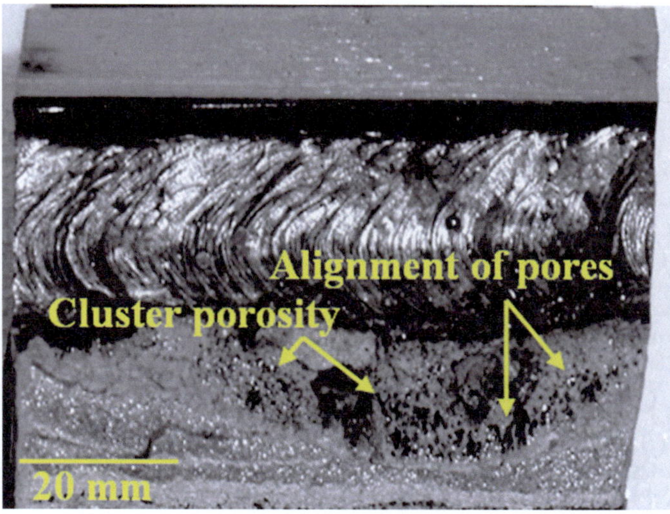


**Fig. 3.5** Typical fracture surface passing through cluster porosity and alignment of pores for a T-welded connection grinded by U-shape grinding and 70 m depth water

## 3.4 Conclusions

A combination of images using SEM and OM were used as a basis scheme to extrapolate the size of 3D pores in T-welded connections with rectangular grinding and filled by wet welding (10 mm grinding depth and 70 mm water depth). It was also shown that the traditional technique to obtain the pore percentage for 2D in cross-sections can be used to relate the mechanical properties of the wet weld beads. However, it is necessary to know the distributions of 3D porosity. The main limitation of the 3D works is that there is no pore size distances data, pore distribution, fraction between each of pore volume, etc. Although there are techniques (μCT and IA) to obtain 3D pores, its main limitation is the resolution (only 2 mm of thickness). It was observed higher pores percentage concentrated in the bottom of the weld bead. It is due to limited space for the gases to escape. In the top weld bead, pores were not found. In this zone there is enough space so that the gases can escape. The main purpose for the porosity is to know the porosity distribution where the mechanical properties of wet welding can be restored and compared with air welding conditions.

**Acknowledgments** The authors would like to thank the Direction de Investigación y Posgrado del Instituto Mexicano del Petréolo and CONACYT for support in this research work.

## References

1. Terán G, Cuamatzi-Meléndez R, Albiter A, Maldonado C, Bracarense AQ (2014) Characterization of the mechanical properties and structural integrity of T-welded connections repaired by grinding and wet welding. Mater Sci Eng A 599:105–115
2. ASM HANDBOOK 6th (2005) Welding, brazing and soldering, ASM International, pp 1010–1016
3. Padilla E, Chawla N, Silva LF, Santos VR, Paciornik S (2013) Image analysis of cracks in the weld metal of a wet welded steel joint by the three dimensional (3D) X-ray microtomography. Mater Charact 83:139–144
4. Pérez-Guerrero F, Liu S (2007) Explaining porosity formation in underwater wet welds. In: Proceeding of the 26th international conference on offshore mechanics and arctic engineering OMAE2007-29696, pp 1–9
5. Pessoa EC, Bracarense AQ, Zica E, Liu S, Perez GF (2006) Porosity variations along multipass underwater processing technology. J Mater Process Technol 179:239–243
6. Paciornik S, Bernthaler T, Santos V, Monteiro M, Mauricio MH, Bracarense A, Pessoa E (2012) Characterization of pores and crack in underwater welds by μCT and digital optical microscopy. In: Proceeding of 1st international conference on 3D material science, Wiley, pp 117–182
7. Terán G, Cuamatzi-Meléndez R, Albiter A (2014) Combination of grinding and wet welding to repair localized cracking in T-welded connections. Mater Sci Forum 793:51–58
8. Ansys (2009) Introduction to ANSYS for release 12.1
9. Puchol RQ, González LP, Scoot AD, Bracarense AQ, Pessoa ECP (2009) Consideraciones termodinámicas entre la formación de poros y la presión hidrostática durante la soldadura subacuática mojada. Soldagem Insp 14(2):161–169
10. Honig EM, Carlson KW (1976) Tensile properties of A514 steel butt joints containing cluster porosity. Weld Res Suppl 103s–107s
11. Jean-Yves B, Maire E, Van-deresse N (2012) 3D characterization and modeling of the influence of porosity on fatigue properties of a cast Al alloy. In: Proceeding of 1st international conference on 3D material science

# Part II
# Characterization of Nanostructured Materials

This part includes four chapters related to the characterization of nanostructured materials, mainly nanoparticles, nanopowders, and nanotubes.

Monometallic and bimetallic nanoparticles are presented in this part. Nanoparticles of Pd, supported on Mg(Al)O supports, were synthesized by vapor phase thermal decomposition method. Small particle sizes and homogenous distribution of Pd nanoparticles on the surface were obtained. PtPd bimetallic nanoparticles were characterized using spherical aberration-corrected scanning transmission electron microscopy. High-angle annular dark-field images of the bimetallic nanoparticles were obtained. The contrast of the images showed that the nanoparticles have an alloy structure, where the atoms of Pd and Pt are found randomly mixed.

On the other hand, multi-walled carbon nanotubes decorated with silver nanoparticles were synthesized by vapor phase impregnation decomposition method. High resolution transmission electron microscopy was employed to investigate structural and morphological features as well as size and distribution of nanoparticles.

Finally, structural and optical properties of pure and Y-doped $ZrO_2$ nanopowders with different Y content sintered by co-precipitation of Zr and Y nitrates were investigated. The decrease of powder grains results in the enhancement of 2.8-eV photoluminescence.

The main techniques used in the characterization of these materials are:

- Transmission electron microscopy (TEM).
- Aberration-corrected scanning transmission electron microscopy (STEM).
- X-ray energy-dispersive spectroscopy (EDS).
- X-ray diffraction (XRD).
- Raman spectroscopy (RS).
- Photoluminescence (PL).
- Cathodoluminescence (CL).

# Chapter 4
# Synthesis and Characterization of Bimetallic Nanoparticles by Cs-Corrected Scanning Transmission Electron Microscopy

**R. Esparza, O. Téllez-Vázquez, A. Ángeles-Pascual, and R. Pérez**

**Abstract** Bimetallic nanoparticles are of interest since they lead to many interesting electrical, chemical, catalytic, and optical properties. They are particularly important in the field of catalysis since they show superior catalytic properties than their monometallic counterparts. The structure of bimetallic nanoparticles depends mainly on the synthesis conditions and the miscibility of the two components. In this work, PtPd bimetallic nanoparticles were synthesized through a chemical reduction method, and characterized using spherical aberration (Cs) corrected scanning transmission electron microscopy (STEM). High-angle annular dark-field (HAADF)-STEM images of the bimetallic nanoparticles were obtained. The contrast of the images shows that the nanoparticles have an alloy structure, where the atoms of Pd and Pt are found randomly mixed, with average particle size of 4.15 nm. Simulated HAADF-STEM images confirm the obtained contrast in the bimetallic nanoparticles due to atomic number of the elements. Beside with the characterization of the nanoparticles, molecular dynamical simulations for the structural and most stable structure of the PtPd bimetallic nanoparticles have also been performed.

**Keywords** Nanoparticles • Synthesis • Bimetallic • Spherical aberration • STEM

## 4.1 Introduction

In the last decade, nanoparticles have drawn increasing attention from researchers due to high variety of properties, compositions, and structures [1]. Nowadays, these particles are used in many new technologies in the field of physics, chemistry,

R. Esparza (✉) • O. Téllez-Vázquez • R. Pérez
Centro de Física Aplicada y Tecnología Avanzada, Universidad Nacional Autónoma de México, Boulevard Juriquilla 3001, Santiago de Querétaro, Querétaro 76230, México
e-mail: resparza@fata.unam.mx

A. Ángeles-Pascual
Departamento de Ingeniería Eléctrica-SEES, CINVESTAV-IPN, Zacatenco, México D.F. 07360, México



R. Pérez Campos et al. (eds.), *Materials Characterization*,
DOI 10.1007/978-3-319-15204-2_4

material science, electrical and chemical engineering [2]. Metallic nanoparticles constitute a new research area because of two main effects: the quantum-size effect and the high surface-to-volume ratio [3].

Bimetallic nanoparticles are the combination of two metals of the nanoscale size range and they are of greater interest compared to monometallic nanoparticles. They are very important in the field of catalysis because they have demonstrated better catalytic properties than their monometallic counterparts [4]. These nanoparticles can be classified into core–shell structures, mixed alloys, and segregated particles [5]. Core–shell nanoparticles consist of a shell of one type of atom surrounding a core of another type of atom. Mixed alloys may be either ordered or random. Random mixed nanoalloys are often termed "alloyed" nanoparticles in the literature. Finally, segregated particles consist of two subclusters which may share a mixed interface or may only have a small number of them.

Platinum and palladium are of interest because they are widely used as catalysts in a number of important reactions [5]. Pd-based electrocatalysts have been extensively investigated because the palladium metal has properties similar to those of platinum, ruthenium, rhodium, osmium, and iridium [6]. In addition, they have demonstrated to be good compounds for use in fuel oxidation and are able to store and release substantial amounts of hydrogen [7]. Bimetallic PtPd nanoparticles are one of the most interesting materials for advanced electrocatalysts [8]. They exhibit a similar catalytic performance and poison resistance compared with Pt during methanol electrooxidation [9]. Thus, bimetallic PtPd electrocatalysts are expected to be a promising substitute for Pt catalysts in direct methanol fuel cells (DMFC) [10].

There are a variety of synthesis methods developed to prepare bimetallic nanoparticles, such as seed-mediated synthesis, polyol synthesis, biological synthesis, hydro/solvothermal synthesis, galvanic replacement reactions, photochemical synthesis, electrochemical synthesis [11], where capping agents, reducing agents, and solvents have been used. However, besides the synthesis process, the characterization of the bimetallic nanoparticles is an important key to understanding their properties.

In the present work, bimetallic PtPd nanoparticles were synthesized by simultaneous reduction of the corresponding precursors. The structural characterization was carried out by scanning transmission electron microscopy (STEM).

## 4.2 Experimental

### *4.2.1 Synthesis of PtPd Alloy Nanoparticles*

The synthesis of the PtPd nanoparticles was carried out by the simultaneous reduction of the corresponding precursors. Briefly, solutions of Pt and Pd ions were prepared by dissolving crystalline chloroplatinic acid hexahydrate ($H_2PtCl_6{\cdot}6H_2O$, 99.99 %) and potassium tetrachloroplatinate ($K_2PdCl_4$, 99.99 %) in hydrochloric acid (HCl) (50 mM in 20 mL of HCl). The solutions were mixed in volumetric ratio 1:1 for the preparation of bimetallic ion mixtures in a three-mouth flask with 10 mL of water de-ionized at vigorous stirring. A solution of PVP [poly(*N*-vinyl-2-pyrrolidone)] (25 mg of PVP of molecular weight 10,000) was added to the metal ion mixtures. To reduce the metal ions, 300 μL of aqueous solution of $NaBH_4$ (50 mM) was added to the mixture solution dropwise at room temperature. A homogeneous colloidal dispersion was formed after the addition of the reducing agent sodium borohydride ($NaBH_4$) in the solution containing metal ions. This colloidal solution was washed using ethanol.

### 4.2.2 *Characterization*

To characterize the bimetallic nanoparticles, copper grids with holey carbon film were prepared with a drop of the corresponding solution. The samples were analyzed using STEM with a Jeol JEM-ARM200F (200 kV) STEM/TEM, equipped with a CEOS Cs-corrector on the illumination system. The probe current used in STEM mode was 23.2 pA using a condenser lens aperture size of 40 μm. High-angle annular dark-field (HAADF)-STEM images were registered using a camera length of 80 mm and a collection angle of 50–180 mrad. Elemental analysis was performed using X-ray energy-dispersive spectroscopy (EDS, Oxford).

HAADF-STEM image simulations have been performed using the QSTEM software package [12] which uses the multislice algorithm [13]. The parameters considered for the simulation correspond to the experimental conditions of the microscope.

### 4.2.3 *Molecular Dynamics*

Molecular dynamics (MD) simulations were used to model the stability of the Pt–Pd nanoparticles. In order to simulate this process, the semiempirical analytic embedded-atom method (EAM) was used. The simulation was performed based on the canonical ensemble (NVT) using the XenoView program. The time step used to perform the simulation was 0.05 femtosecond (fs). Periodic boundary conditions were applied in all the directions of the simulation cell. Cubo-octahedral structures with 490 atoms were the models of the nanoparticles used in the simulation.

## 4.3 Results and Discussion

Characterize the structure of the bimetallic PtPd nanoparticles by traditional high-resolution transmission electron microscopy (HR-TEM) imaging is difficult because both components exhibit high contrast. So that, high-angle annular dark-field scanning transmission electron microscopy (HAADF-STEM) is the best technique to analyze the structure of the bimetallic nanoparticles [14], given that the intensity

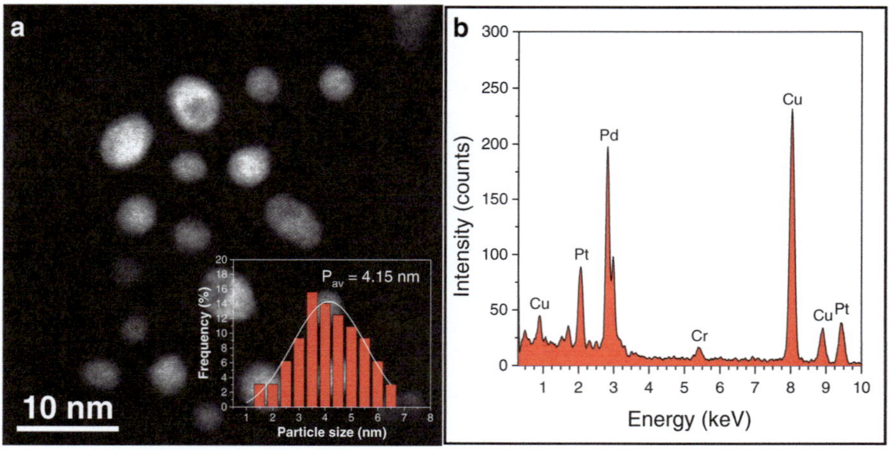


**Fig. 4.1** (**a**) HAADF-STEM image and its corresponding particle size distribution of PtPd nanoparticles, and (**b**) EDS spectrum confirming the presence of Pt and Pd

contrast of the images is directly proportional to the density and thickness of the specimen and proportional to atomic number of the sample ($Z^{3/2}$) [15]. Figure 4.1a shows a low magnification HAADF-STEM image of the PtPd bimetallic nanoparticles. As can be observed in the image, the nanoparticles showed a particle size distribution almost homogeneous, with an average particle size of 4.15 nm (inset). It is very important to mention that coalescence of the nanoparticles was not observed, and the contrast of the image is related with homogeneous alloy particles [16], in others words, variation in contrast which is characteristic of core–shell structures is not observed [17]. Figure 4.1b shows the energy-dispersive X-ray (EDS) spectra acquired from the bimetallic nanoparticles showing both Pt and Pd peaks suggesting mixing of the two elements in the samples.

To characterize the structure of the nanoparticles, high-resolution HAADF-STEM images were analyzed. Figure 4.2a shows a high-resolution HAADF-STEM image of a PtPd alloy nanoparticle, where the atomic resolution clearly can be seen. Measuring the reciprocal lattice vectors using the fast Fourier transform (FFT) is a common method in HR-TEM. Using the FFT, one can obtain much more precise measurements. Figure 4.2b shows the FFT of the PtPd nanoparticle where 0.2247, 0.1943, and 0.1340 nm *d*-spacings were obtained, such *d*-spacings corresponding to (1−11), (0 0 2), and (2–20) crystalline planes, respectively, of the face centered cubic (fcc) structure of the PtPd with a lattice parameter $a_0 = 0.3896$ nm (JCPDS card no. 65-6418) and the particle is oriented in the [1 1 0] zone axis. As previously mentioned, the intensity of the HAADF-STEM images is directly proportional to the density and thickness of the specimen and proportional to atomic number of the sample ($Z^{3/2}$) [15]. From the image it is clear that the differences in contrast are due to the difference in atomic numbers between the two elements. To visualize a better contrast, Apply_CLUT (color look-up table) script was used [18]. It is useful for highlighting subtle intensity variations, which are

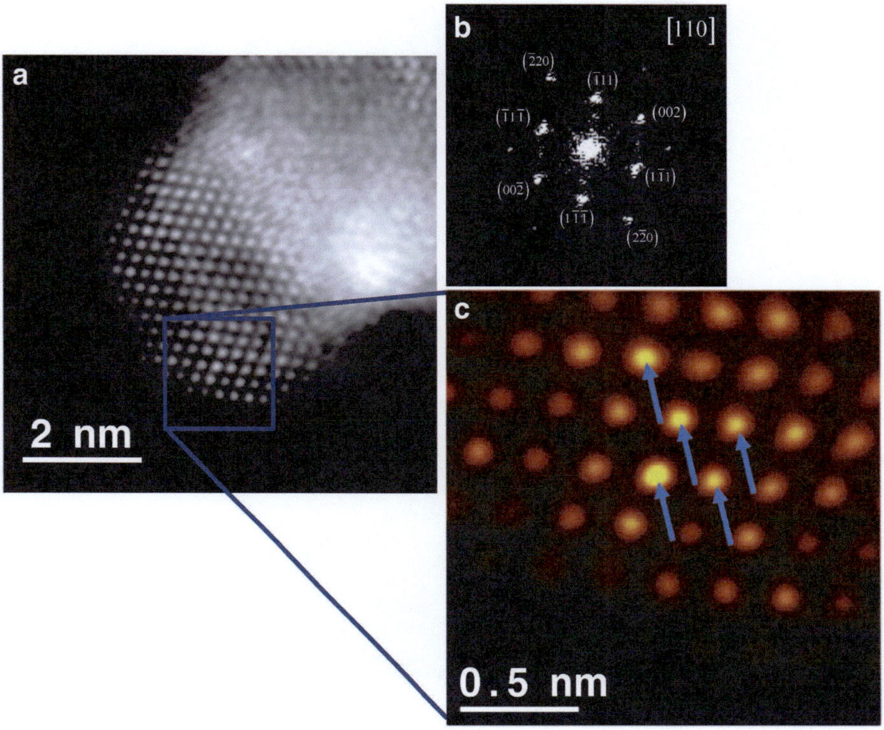


**Fig. 4.2** (**a**) High-resolution HAADF-STEM image of the PtPd nanoparticle, (**b**) FFT shows the corresponding lattice indexing of the PtPd alloy, and (**c**) CLUT image of the inset HAADF-STEM image

much more evident in color than grayscale. Figure 4.2c shows the CLUT image from a region of the HAADF-STEM image of the PtPd nanoparticle. From the image, clearly can be observed the difference in contrast, the arrows show the atomic columns with high contrast, which they can be related with a high Pt content. These images confirm that the PtPd bimetallic nanoparticles have an alloy structure where the Pt and Pd atoms are random localized.

Figure 4.3a shows a high-resolution HAADF-STEM image of PtPd nanoparticle oriented in the [1 0 1] zone axis. This image shows clearly the difference in intensity contrast, which is related with the atomic number of the Pt and Pd, as previously mentioned, to obtain a better calculation about the composition of the atomic columns, the intensity of the signal from the atomic columns in the experimental image was extracted and plotted as a color map representing integrated intensity of the atoms. Figure 4.3b shows the map representing the integrated intensity of the atomic columns plotted using Absolute Integrator v1.2 program, which is a useful code to quantitative analysis of HAADF-STEM images, for the calculation and visualization of absolute scattering from atomic-resolution images. The color scale of the mapping image is related with the Pt content. As can be observed, the dark red points are the

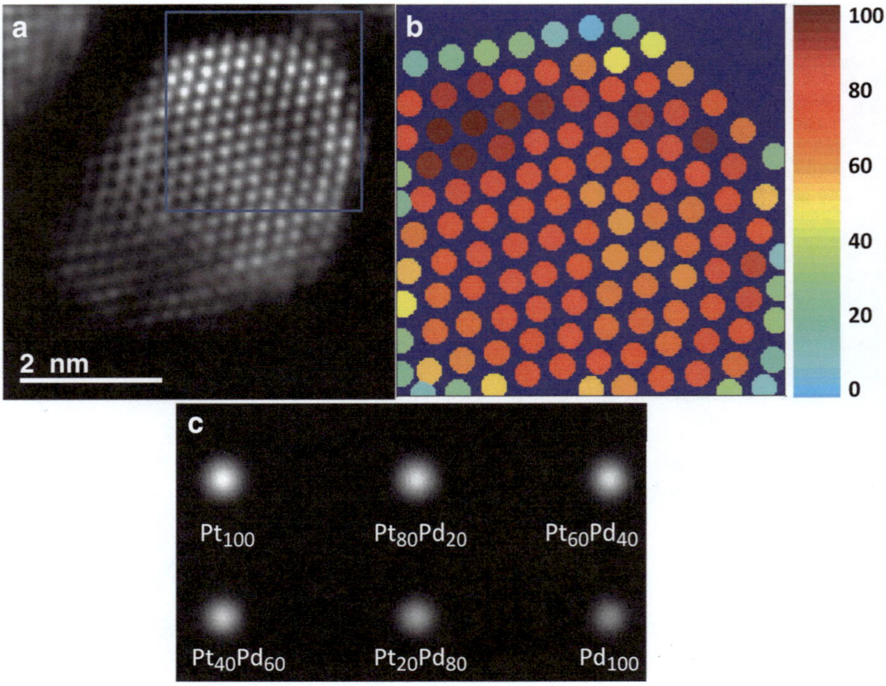


**Fig. 4.3** (**a**) Experimental HAADF-STEM image of PtPd nanoparticle, (**b**) map representing integrated intensity of the atomic columns, and (**c**) simulated HAADF-STEM contrast of the atomic columns with different composition of Pt and Pd

atomic columns with higher Pt content; otherwise, the yellow and green points are the atomic columns with less Pt content or more Pd content. Figure 4.3c shows a simulated HAADF-STEM image with different compositions in atomic percent of atoms. The calculations of the HAADF-STEM contrast of different compositions were performed with the same number of atoms. It is clear that the column with 100 at% Pt has a higher contrast; otherwise, column with 100 at% Pd has a less contrast. HAADF-STEM images are sensitive to atomic number of the sample, the electron scattering of elements with high atomic number (heavy elements) makes those atomic columns display more contrast and with a larger diameter compared to atomic columns of light elements [19], as happens in the case of Pt and Pd.

Figure 4.4 shows the energy per atom as a function of temperature obtained by the molecular dynamics (MD) simulations. The jump in the curve indicates the solid–liquid phase transition (latent heat of fusion). The transition temperature about 1,400 K for $Pt_{25}Pd_{75}$ is clearly shown, this temperature is increased with the Pt content, i.e. for $Pt_{50}Pd_{50}$ the transition temperature is about 1,500 K. If the Pt content is increased, the transition temperature increases up to the latent heat of fusion of the Pt, which is higher than 1,700 K; this temperature is not shown in the graphic. Previous works have shown that the transition temperature or melting points to PtPd alloy is approximately 1,600 K, which is in agreement with the obtained results [20].

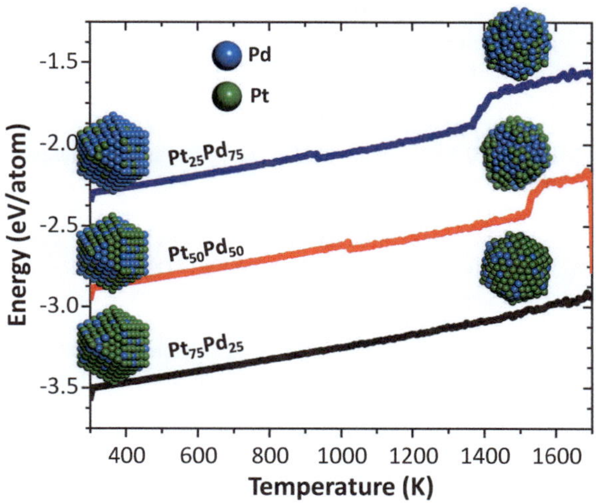


**Fig. 4.4** Correlation of the total energies with the temperature of the PtPd structures

## 4.4 Conclusions

PtPd bimetallic nanoparticles were characterized by using Cs-corrected scanning transmission electron microscopy. The synthesis of the bimetallic nanoparticles was carried out by simultaneous chemical process, obtaining an average particle size of 4.15 nm. Bimetallic nanoparticles had an alloy structure with random Pt and Pd atom positions. HAADF-STEM images allowed to identify the atomic columns with higher Pt content and similarly, atomic columns with higher Pd content; it is due to intensity image contrast result of the atomic number of the components. Molecular dynamics simulations were employed to examine the structural and stability behavior of PtPd bimetallic nanoparticles with alloy structure. It was found that the solid–liquid phase transition temperature depends on the composition of the nanoparticles, with higher melting point alloys with high content of Pt.

**Acknowledgments** The authors would like to acknowledge the PAPIIT-DGAPA for financial support with grant IN110113. Also, the authors acknowledge the Laboratorio Avanzado de Nanoscopia Electrónica "LANE" at the CINVESTAV-Zacatenco.

## References

1. Schmid G (2008) General features of metal nanoparticles physics and chemistry. In: Corain B, Schmid G, Toshima N (eds) Metal nanoclusters in catalysis and materials science: the issue of size control. Elsevier, Amsterdam
2. Toshima N, Yonezawa T (1998) Bimetallic nanoparticles – novel materials for chemical and physical applications. New J Chem 22:1179–1201
3. Roduner E (2006) Size matters: why nanomaterials are different. Chem Soc Rev 35:583–592

4. Devarajan S, Bera P, Sampath S (2005) Bimetallic nanoparticles: a single step synthesis, stabilization, and characterization of Au–Ag, Au–Pd, and Au–Pt in sol–gel derived silicates. J Colloid Interface Sci 290:117–129
5. Ferrando R, Jellinek J, Johnston RL (2008) Nanoalloys: from theory to applications of alloy clusters and nanoparticles. Chem Rev 108:845–910
6. Campbell FW, Compton RG (2010) The use of nanoparticles in electroanalysis: an updated review. Anal Bioanal Chem 396:241–259
7. Langhammer C, Zhdanov VP, Zoric I, Kasemo B (2010) Size-dependent kinetics of hydriding and dehydriding of Pd nanoparticles. Phys Rev Lett 104:135502
8. Chen X, Cai Z, Chen X, Oyama M (2014) Synthesis of bimetallic PtPd nanocubes on graphene with N,N-dimethylformamide and their direct use for methanol electrocatalytic oxidation. Carbon 66:387–394
9. Yang X, Yang Q, Xu J, Lee C (2012) Bimetallic PtPd nanoparticles on Nafion-graphene film as catalyst for ethanol electrooxidation. J Mater Chem 22:8057–8062
10. Xia T, Shen H, Chang G, Zhang Y, Shu H, Oyama M, He Y (2014) Facile and rapid synthesis of ultrafine PtPd bimetallic nanoparticles and their high performance toward methanol electrooxidation. J Nanomater 1:1–7
11. Sajanlal PR, Sreeprasad TS, Samal AK, Pradeep T (2011) Anisotropic nanomaterials: structure, growth, assembly, and functions. Nano Rev 2:5883
12. Koch C (2002) Determination of core structure periodicity and point defect density along dislocations. PhD thesis, Arizona State University, USA
13. Cowley JM, Moodie AF (1957) The scattering of electrons by atoms and crystals. I. A new theoretical approach. Acta Cryst 10:609–619
14. Esparza R, García-Ruiz AF, Velázquez Salazar JJ, Pérez R, José-Yacamán M (2013) Structural characterization of Pt–Pd core–shell nanoparticles by Cs-corrected STEM. J Nanopart Res 15:1342
15. Nellist PD, Pennycook SJ (2000) The principles and interpretation of annular dark-field Z-contrast imaging. Adv Imag Electron Phys 113:147–203
16. Dimitratos N, Edwards JK, Kiely CJ, Hutchings GJ (2012) Gold catalysis: helping create a sustainable future. Appl Petrochem Res 2:7–14
17. Esparza R, Téllez-Vázquez O, Rodríguez-Ortiz G, Ángeles-Pascual A, Velumani S, Pérez R (2014) Atomic structure characterization of Au–Pd bimetallic nanoparticles by aberration-corrected scanning transmission electron microscopy. J Phys Chem C 118:22383–22388
18. Mitchell D, Schaffer B (2005) Scripting-customised microscopy tools for Digital Micrograph™. Ultramicroscopy 103:319–332
19. Mayoral A, Esparza R, Deepak FL, Casillas G, Mejía-Rosales S, Ponce A, José-Yacamán M (2011) Study of nanoparticles at UTSA: one year of using the first ARM-200F installed in USA. JEOL News 46:1–5
20. Huang R, Wen YH, Zhu ZZ, Sun SG (2012) Pt–Pd bimetallic catalysts: structural and thermal stabilities of core–shell and alloyed nanoparticles. J Phys Chem C 116:8664–8671

# Chapter 5
# Application of Silver Decorated Carbon Nanotubes for Environmental Ozone Sensing

**S. Capula Colindres, G. Terán, V. Garibay Febles, L.A. Villa Vargas, and J.R. Vargas García**

**Abstract** This work reports a gas sensor development based on multi-walled carbon nanotubes (MWCNTs) decorated with silver nanoparticles and fabricated by vapor phase impregnation decomposition method (VPID). The sensitive material was deposited on a substrate type resistor by drop coating. The gas sensing behavior of multi-walled carbon nanotubes-silver (MWCNTs-Ag) was evaluated using a calibrated commercial ozone generator. After a 1 min ozone exposure, MWCNTs-Ag showed electric resistance changes at different operation temperatures. The sensor was evaluated at 300 ppb and lower than 100 ppb. Results show that the MWCNTs-Ag exhibits the best sensitivity response at room temperature. The sensor response depends on the ozone concentration as the electric resistance increases with higher gas concentration, reporting p-type characteristics upon gas exposures. Raman spectroscopy was used to study the quality of MWCNTs after being subjected to an oxidative process. High resolution transmission electron microscopy (HRTEM) was employed to investigate structural and morphological features as well as size and distribution of nanoparticles (NPs). The responsiveness of MWCNTs decorated with different NPs, i.e. platinum (Pt), palladium (Pd), and silver (Ag) is compared at room temperature, 120 and 200 °C. The best performance was from MWCNTs-Ag at room temperature, while at 120 and 200 °C MWCNTs-Pd reported the best behavior.

**Keywords** Carbon nanotubes • Silver • Ozone • Sensor • Vapor phase impregnation decomposition method

---

S. Capula Colindres (✉) • G. Terán • V. Garibay Febles
Mexican Petroleum Institute, Eje Central Lázaro Cárdenas 152, 07730 México City, México
e-mail: selenecapula@gmail.com; gerardoteranm@gmail.com; vgaribay@imp.mx

L.A. Villa Vargas
Centro de Investigación en Computación, National Polytechnic Institute, Av. Juan de Dios Batiz s/n Col. Nueva Industrial Vallejo, 07738 México City, México
e-mail: lvilla@cic.ipn.mx

J.R. Vargas García
Department of Metallurgical Engineering, National Polytechnic Institute, 07300 México City, México
e-mail: rvargasga@ipn.mx

 
R. Pérez Campos et al. (eds.), *Materials Characterization*,
DOI 10.1007/978-3-319-15204-2_5

## 5.1 Introduction

In the last years, gas sensors based on carbon nanotubes (CNTs) have attracted extensive interest due to the small size of sensitive elements, plus the fact that they are effective at room temperature and exhibit a very fast response time, low power consuming electronics, high response, and relatively low fabrications cost [1, 2]. This type of sensor has been employed for monitoring chemical molecules such as nitrogen dioxide [3], ammonia [4], ozone [5, 6], methanol, and acetone [2]; for biological, public security, agriculture, and other fields. The U.S. environmental protection agency determined that the air quality standard for concentration of ozone must be lower than 75 ppb. Higher ozone concentration levels can damage human health and vegetation in forested ecosystems.

CNTs are very sensitive, even in small quantity; they can easily change their conductivity in the presence of different gases. In order to improve their selectivity it is necessary to modify their surface. Different methods have been developed to both adjust the electronic properties of nanotubes and increase their selectivity towards the determination of specific molecules using among others, organic polymers, functional groups, metal nanoparticles, and nanostructured oxides [1, 2, 7, 8]. Incorporation of metal nanoparticles (NPs) on CNTs is an attractive way to improve the adsorption mechanism of the gas because the NPs reveal interesting chemical, electronic, and physical properties plus the fact that they can work in harsh environments [1, 2]. Nowadays, sensors based on single wall carbon nanotubes (SWCNTs) [5, 6, 9] and MWCNTs with metallic NPs [10, 11] have been studied to detect ozone. Sensors which have SWCNTs as sensitive material reported high responsiveness to ozone, reaching a maximum of 14.1 % (1 ppm) and a minimum of 3 % (21.5 ppb), nevertheless, the response time was between 10 and 40 min. On the other hand, MWCNTs decorated with Pd or Pt NPs exhibited efficient sensitivity and a much shorter response time (less than 1 min). By combining properties of metal nanoparticles and CNTs, it is possible to achieve a satisfactory and efficient sensor response. However, the sensitivity and selectivity of the sensor depends upon the interaction between the CNTs and metal NPs.

This work reports a gas sensor based on MWCNTs which were decorated with Ag nanoparticles. Ag NPs were incorporated by VPID method. The sensitive material was deposited on a substrate with resistor configuration by drop coating. The gas sensing behavior of MWCNTs-Ag was compared with the response of MWCNTs decorated by Pd and Pt NPs when exposed to ozone gas at different temperatures and concentrations. The interaction between MWCNTs and NPs plays a crucial role when trying to improve the efficiency of the sensor response.

## 5.2 Experimental Work

MWCNTs decorated with Ag NPs were used as sensitive material. The sensing platform consists of two comb-like electrodes upon which the sensitive material solution was deposited. The MWCNTs used in this work were synthesized by chemical vapor deposition (CVD) of ferrocene and toluene under Ar atmosphere as previously reported [10]. Raw nanotubes were treated by an oxidative process in order to eliminate impurities and to reactivate the walls of the nanotubes. This method is described in a previous paper [10]. Silver nanoparticles (NPs Ag) were incorporated by the VPID method, as said before [10–12]. Silver acetate ($C_2H_3AgO_2$) and MWCNTs were mechanically mixed for 15 min. The sample was introduced inside a horizontal quartz-tube reactor and heated to 180 °C and then to 400 °C; in every heating stage, the same parameters were employed (10 min at a pressure of 66.6 kPa [11, 12]). The sensitive material was characterized by high resolution transmission electron microscopy (HRTEM) in a Tecnai G2 F30 microscope, with the voltage set at 300 kV. The sensitive material was deposited on the device by drop coating and dried at 200 °C for 2 h. The sensor was placed inside a sealed stainless steel test cell. Test gas concentration was controlled with a calibrated commercial ozone generator (InDevR 2B Technologies model 306) and monitored by a Keithley 6430 digital multimeter.

## 5.3 Results and Discussion

The resulting raw CNTs are shown in Fig. 5.1a, which revealed their multiwalled nature. The outer average diameter of the nanotubes was 52.7 nm and several hundred microns long. A HRTEM image of individual MWCNT, after being exposed to acid-based chemical modification is presented in Fig. 5.1b. Comparing the morphology of the as-prepared MWCNTs and nanotubes which were exposed to the acid treatment, important morphological changes are apparent. After acid treatment, MWCNTs present sidewall alteration, even becoming amorphous. Furthermore, the amount of nanotubes decreases; this behavior is attributed to the aggressive nature of the oxidative treatment, because the process tends to destroy SWCNTs and the nanotubes with few walls [13].

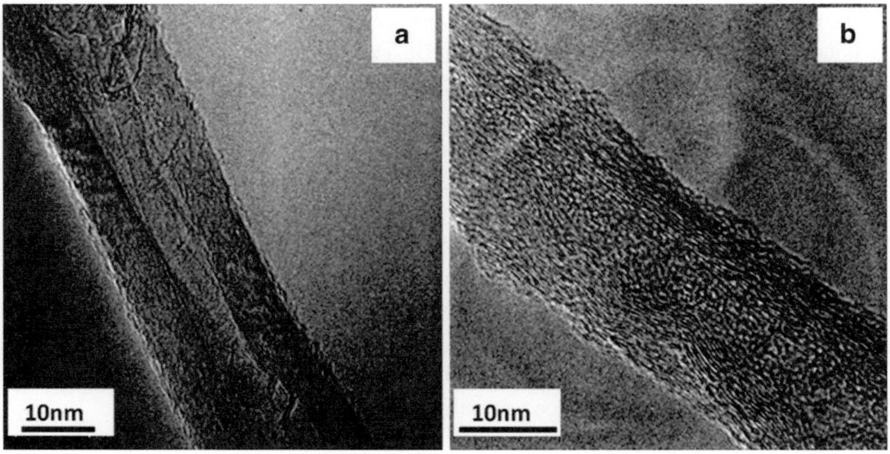


**Fig. 5.1** HRTEM images of single MWCNTs; (**a**) as prepared and (**b**) after acid treatment

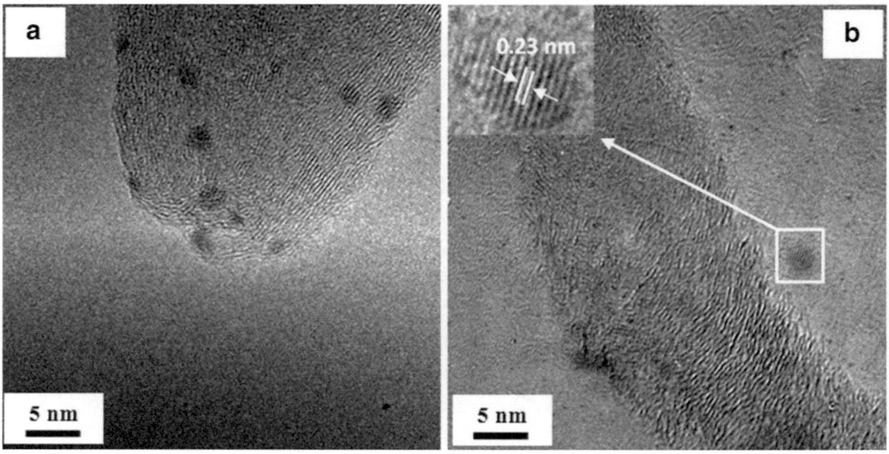


**Fig. 5.2** HRTEM images of individual MWCNTs decorated with Ag NPs; (**a**) distribution of Ag NPs and (**b**) an individual Ag NPs which has a lattice fringe of 0.23 nm, according to the target JCPDS card 03-0921

Figure 5.2a, b shows HRTEM images of typical samples of MWCNTs decorated with Ag NPs; the size of these nanoparticles was in the 1–5 nm range. NPs present uniform decoration on both the outside walls and the tip of MWCNTs (Fig. 5.2a). Figure 5.2b shows a nanotube; the insert in it shows a single particle, which has reported lattice fringe of 0.23 nm, corresponding to the (1 1 1) plane of Ag crystal (JCPDS 03-0921). The efficient distribution of nanoparticles is due to the introduction of oxygenated groups such as carboxylic group by acid treatment. Balasubramanian and Burghard [7] reported that the presence of the carboxyl groups on CNTs represents helpful sites for further modifications, for example dendrimers, nucleic acids, enzymes, semiconductor, and metal nanoparticles. Also, the acid process is useful for removing impurities present in raw nanotubes.

The Raman spectra of raw and acid treated MWCNTs excited with a 633 nm laser line are shown in Fig. 5.3. Raman spectra showed four strong scattering peaks of CNTs: D band, G band, G′ band, and G+D band [10]. The bands were observed about 1,333, ~1,576, 2,662, and 2,919 $cm^{-1}$. By comparison of the two spectra we can observe: (1) a high intensity D band on the sample after treatment. This behavior suggests the presence of defects on the nanotube walls introduced by the oxidative process [14], (2) raw MWCNTs showed a mode at 1,462 $cm^{-1}$. This mode has been reported before by Chiu group [15]; they explained the presence of this band due to the resonance C=C bond on the $C_{60}$ of the CNT of its net structure and (3) the sample treatment caused a shifting of band G to higher frequencies. The difference could be attributed to a possible intercalation of foreign atoms from $HNO_3$ molecules into the CNTs structure [16].

The typical responsiveness of the MWCNTs-Ag gas sensor in terms of percentage in presence of $O_3$ is shown in Fig. 5.4a. The sensitivity of sensor was measured by the shift of the electric resistance normalized:

$$S(\%) = \left(R_g - R_0\right) / R_0 \times 100 \qquad (5.1)$$

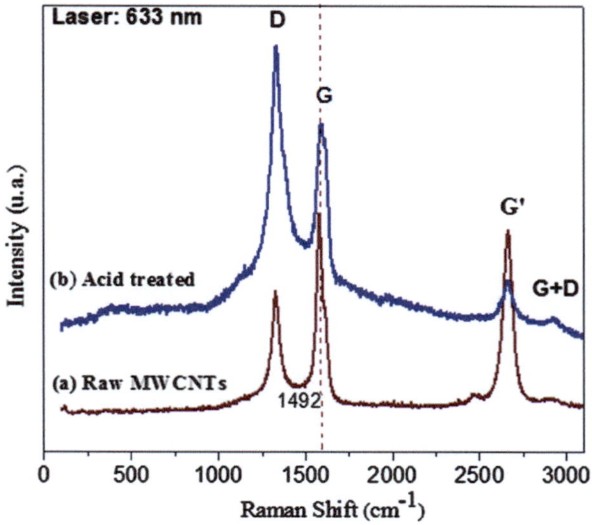


**Fig. 5.3** Raman spectra of MWCNTs; (**a**) raw sample and (**b**) acid treated sample

where $R_g$ is the resistance of the sensor when the sensitive material is exposed to the presence of a gas and air mixture concentration and $R_0$ stands for the air resistance alone [10]. The MWCNTs-Ag showed an efficient response to $O_3$ in the range from room temperature to 200 °C at different concentrations (20–300 ppb), where the sensor showed the best response at room temperature. The studied sensor reported a response time of less than 1 min at all temperatures while recovery time and intensity of signal increased when the ozone concentration increased. The sensor presented a p-type semiconductor behavior. Figure 5.4b–d presents a comparison of sensor response of MWCNTs decorated with different metal nanoparticles (Pd, Pt, and Ag) evaluated under the same condition. Incorporation of Pt and Pd NPs was the same protocol that Ag NPs by VPID method. At room temperature, MWCNTs-Ag reported the best responsiveness (Fig. 5.4b), while at 120 and 200 °C MWCNTs-Pt showed the best behavior (Fig. 5.4c, d).

Comparing size and NPs dispersion of the three sensitive materials tested, we found that MWCNTs-Ag reported an NPs size range of 1–5 nm and uniform dispersion, MWCNTs-Pd [10] showed an efficient and homogeneous dispersion of Pd NPs on the surface of CNTs but the average size of NPs was about 10.8 nm and finally, MWCNTs-Pt [11] reported a nanoparticle size of about 1.47 nm and heterogeneous dispersion. At room temperature, the sensor that reported NPs moderate size and efficient dispersion had the best sensitivity to $O_3$, meanwhile the sensor that presented a small size and NPs heterogeneous dispersion reported a low responsiveness. According to the results in this work, the responsiveness of the sensor may be attributed to the interaction between NTCs and nanoparticles. Additionally, the efficient sensor response of MWCNTs-Pt at 120 and 200 °C is influenced by heating cycles. Bonet et al. [17] claimed that the catalytic properties of NPs depend on their average size and in some application is required uniform dispersion. Chen and Lu [18] reported Ag NPs-CNT hybrid structure; Ag nanoparticles were incorporated by

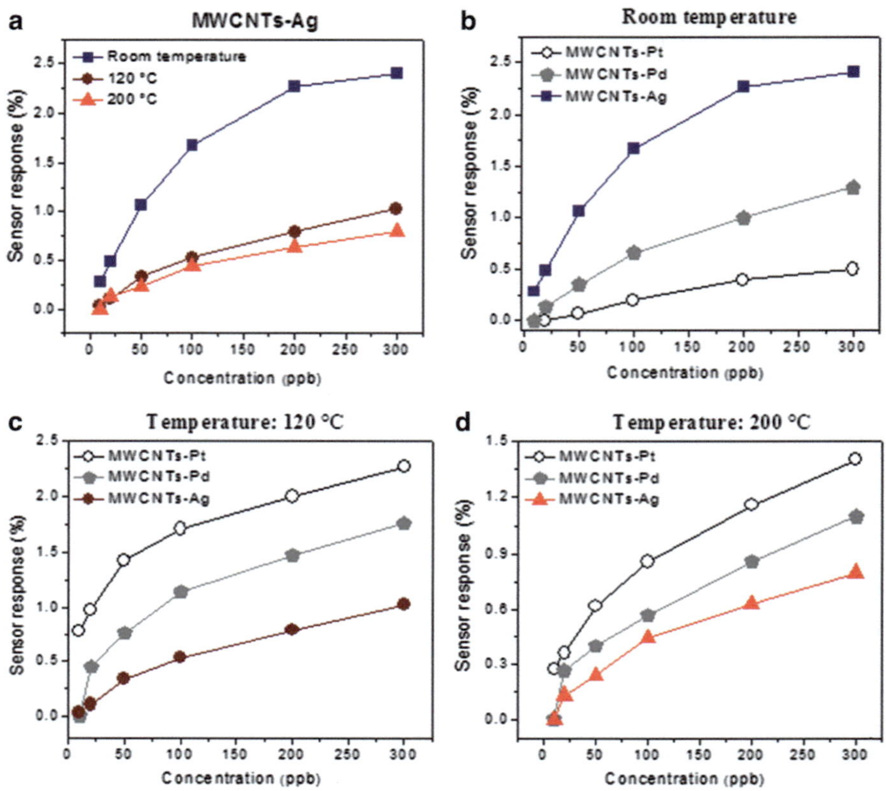


**Fig. 5.4** Sensor response curves based on MWCNTs; (**a**) MWCNTs-Ag at different temperatures and concentration, (**b**) MWCNTs-metal at room temperature, (**c**) MWCNTs-metal at 120 °C and (**d**) MWCNTs-metal at 200 °C

an electrostatic force directed assembly (ESFDA). The hybrid was heated at 100, 200, and 300 °C for 1 h. After heating, Ag NPs presented an increase in the average diameter and a decrease in the density of NPs on CNT surfaces. Furthermore, the same author claimed that at relatively low temperatures (100–200 °C), size particle increase due to the coalescence of the smaller particles to form larger ones, due to the weak interaction between NPs and CNTs. Additionally, we reported in previous work [10] that NPs of noble metals are highly mobile and they migrate to the surface and agglomerate into larger particles due to weak interaction. According to the research made by Durgun et al. [19] and Zhang et al. [20] electrical characteristics of metal-nanotube system are sensitive to the metal type. They reported an agglomeration of gold (Au) nanoparticles on SWCNTs surfaces due to the very weak interaction (Van der Waals) between Au atoms with SWCNTs. This behavior is attributed to both: an insufficient affinity for carbon and to low binding energy of Au element (0.5 eV) since it has no $d$-vacancies ($5d^{10}\ 6s^1$). On the contrary, when the metal elements show high binding energy and it exhibits fewer $d$ electrons ($3d^2\ 4s^2$) such as

Titanium (Ti) coating appears very uniform on the nanotube sidewalls due to the interactions between metal NPs and CNT is a chemical bond. In case of Ag NPs, the coalescence of particles seems to be related to the low binding energy (0.2 eV) of Ag, which in turn results in a weak interaction between Ag atoms and CTNs.

## 5.4 Conclusions

In conclusion, we manufactured a gas sensor based on MWCNTs. In order to improve their sensitivity and selectivity, the carbon nanotubes were decorated with Ag nanoparticles. The sensor reported efficient responsiveness at room temperature. Sensor response was compared with other MWCNTs sensors decorated with Pt and Pd NPs at different temperatures. The best response was that from MWCNTs-Ag at room temperature while at relatively high temperatures the best behavior was that from MWCNTs decorated with Pt. The interaction between CNTs and NPs plays a crucial role in the development of an $O_3$ detecting sensor.

**Acknowledgments** The authors wish to thank Dr. S. Mendoza Acevedo, Centro de Investigación en Computación for insightful discussions and Dr. Cervantes, Universidad Iberoamericana for assistance with sample synthesis. This work was financially supported by CONACyT and IMP to work at Aix-Marseille University, IMP2NP of France to perform the testing program.

## References

1. Zhang W, Zhang W (2009) Carbon nanotubes as active components for gas sensor. J Sensors 2009:16. doi:10.1155/2009/160698
2. Zhang T, Mubeen S, Myung N, Deshausses M (2008) Recent progress in carbon nanotube-based gas sensors. Nanotechnology 19:1–1
3. Cantalini C, Valentini L, Lozzi L, Armentano I, Kenny J, Santucci S (2003) $NO_2$ gas sensitivity of carbon nanotubes obtained by plasma enhanced chemical vapor deposition. Sens Actuators B 93:333–337
4. Suehiro J, Zhou G, Hara M (2003) Fabrication of a carbon nanotube-based gas sensor using dielectrophoresis and its application for ammonia detection by impedance spectroscopy. J Phys D Appl Phys 36:109–114
5. Park Y, Lee J, Choi J, Bae G, Ju B (2009) Development of an ozone gas sensor using single walled carbon nanotubes. Sens Actuators B 140:407–411
6. Wongwiriyapan W, Honda D, Konishi H, Mizuta T, Kuno T, Ito T, Maekawa T, Suzuki K, Ishikawa H, Murakami T (2006) Ultrasensitive ozone detection using single walled carbon nanotube networks. Jpn J Appl Phys 45:3669–3671
7. Balasubramanian K, Burghard M (2005) Chemically functionalized carbon nanotubes. Small 2:2679–2694
8. Georgakilas V, Gournis D, Tzitzios V, Pasquato L, Guldi D, Prato M (2007) Decorating carbon nanotubes with metal or semiconductor nanoparticles. J Mater Chem 17:2679–2694
9. Ghaddab B, Sanchez J, Berger F (2012) Detection of $O_3$ and $NH_3$ using hybrid tin dioxide/carbon nanotubes sensors: influence of materials and processing on sensor's sensitivity. Sens Actuators B: Chem 170:67–74

10. Capula S, Aguir K, Cervantes F, Villa L, Moncayo J, Garibay V (2014) Ozone sensing based on palladium decorated carbon nanotubes. Sensors 14:6806–6818
11. Capula S, Aguir K, Cervantes F, Villa L, Garibay V (2014) Carbon nanotubes functionalized by nanoparticles of platinum. Mater Sci Forum 793:45–50
12. Encarnación C, Vargas J, Toledo J, Cortes M, Angeles C (2010) Pt nanoparticles on titania nanotubes prepared by vapor-phase impregnation–decomposition method. J Alloy Compd 495:458–461
13. Park T, Barnerjee S, Hemrai T, Wong S (2006) Purification strategies and purity visualization techniques for single walled carbon nanotubes. J Mater Chem 16:141–154
14. Watts P, Mureau N, Tang Z, Miyajima Y, Carey D, Silva S (2007) The importance of oxygen containing defects on carbon nanotubes for the detection of polar and non-polar vapors through hydrogen bond formation. Nanotechnology 18:175701
15. Chiu W, Chang Y (2008) Chemical modification of multiwalled carbon nanotubes with the liquid phase method. J Appl Polym Sci 107:1655–1660
16. Martinez M, Callejas M, Benito A, Cochet M, Seeger T, Ansón A, Schreiber J, Gordon C, Marhic C, Chauvert O, Fierro J, Maser W (2003) Sensitivity of single wall carbon nanotubes to oxidative processing: structural modification, intercalation and functionalization. Carbon 41:2247–2256
17. Bonet F, Delams S, Grugeon S, Herrera R, Silvert P, Tekaia K (1999) Synthesis of monodisperse Au, Pt, Pd, Ru, and Ir nanoparticles in ethylene glycol. Nano Struct Mater 11:1277–1284
18. Chen J, Lu G (2010) Carbon nanotube-nanoparticles hybrid structures. In: Marulanda JM (ed) Carbon nanotubes. http://www.intechopen.com/books/carbon-nanotubes/carbon-nanotube-nanoparticle-hybrid-structures
19. Durgun E, Dag S, Bagci V, Gulseren O, Yildirim T, Ciraci S (2003) Systematic study of adsorption of single atoms on a carbon nanotube. Phys Rev B 67:201401–201404
20. Zhang Y, Franklin W, Chen R, Dai H (2000) Metal coating on suspended carbon nanotubes and its implication to metal-tube interaction. Chem Phys Lett 24:35–41

# Chapter 6
# Spectroscopic and Structural Characterization of Pd Nanoparticles Supported on Hydrotalcite-Like Materials

**Miguel Angel de la Rosa-Guzmán, María de Jesús Martínez-Ortiz, Nancy Castillo, Carlos Alberto Ramírez-Salazar, Andrés Méndez-Ceja, and Juan Esteban Canseco-Morales**

**Abstract** Nanoparticles of Pd with 0.5 wt% loading, supported on mixed oxides coming from Mg/Al layered double hydroxides, were synthesized by vapor phase thermal decomposition method. Pd/Mg(Al)O metallic nanoparticles were characterized by spectroscopic and structural techniques by using X-ray diffraction (XRD), infrared spectroscopy (FTIR), scanning electron microscopy (SEM), high resolution transmission electron microscopy (HRTEM), and $N_2$ sorption experiments. Small particle sizes and homogenous distribution of Pd nanoparticles on the surface of Mg(Al)O supports were obtained by using this method of metal incorporation.

**Keywords** Nanoparticles • X-ray diffraction (XRD) • Infrared spectroscopic (FTIR) • Scanning electron microscopy (SEM) • High resolution transmission electron microscopy (HRTEM)

## 6.1 Introduction

Layered double hydroxides (LDH), usually known as hydrotalcite like-materials (HDT), belong to the anionic clays group. They have a general formula $[M^{II}_{1-x}M^{III}_{x}(OH)_2]^{x+}[A^{n-}_{x/n}]\cdot mH_2O$, similar to the brucite-like layers $Mg(OH)_2$. The brucite is constituted by octahedrons which share their edges with $OH^-$ groups, but when some of the Mg atoms are replaced by Al atoms, this causes an excess of positive charge which needs to be compensated by anionic species and water molecules, both placed in the interlayered space [1–3]. HDT materials have been very useful in

M.A. de la Rosa-Guzmán • M.J. Martínez-Ortiz (✉) • N. Castillo
C.A. Ramírez-Salazar • A. Méndez-Ceja • J.E. Canseco-Morales
Instituto Politécnico Nacional-ESIQIE, Laboratorio de Investigación en Materiales Porosos, Catálisis Ambiental y Química Fina, UPALM Edif. 7 P. B., Zacatenco, C. P. 07738 México, D.F., México
e-mail: mjmartinez@ipn.mx



R. Pérez Campos et al. (eds.), *Materials Characterization*,
DOI 10.1007/978-3-319-15204-2_6

different applications as catalysts and supports of multifunctional catalysts due to their own acid–base properties, considerable surface area, and memory effect, among others [4–6]. In addition, the impregnation of metallic phases by vapor phase thermal decomposition method has been used for different purposes with the aim to obtain very small and homogenous metallic particle sizes [7–10]. Thus, the goal of this work is to investigate the metal dispersion degree and the particle size of the materials to be applied as catalysts in the selective hydrogenation reactions.

## 6.2 Experimental

Mg/Al HDT like-materials with different $M^{II}/M^{III}$ molar ratios were prepared by the coprecipitation method, at constant pH (~10) from suitable amounts of $Mg(NO_3)_2{\cdot}6H_2O$ and $Al(NO_3)_3{\cdot}6H_2O$ (Aldrich), with a 2.0 M solution of NaOH (Aldrich). The addition of the alkaline solution and pH were controlled by a pH-STAT Titrando apparatus (Metrohm, Switzerland). The resulting suspension was stirred overnight at 353 K. The solid was separated by centrifugation and rinsed thoroughly with distilled water and it was dried overnight at 353 K. These HDT samples were labelled as HDT2, HDT3, and HDT4 depending on the Mg/Al molar ratio used during the synthesis. The solids then followed a calcination process in air flow at 823 K for 3 h (3 K $min^{-1}$) to obtain the corresponding Mg(Al)O mixed oxides. After this thermal treatment, the catalyst supports were labelled as HDT2TT, HDT3TT, and HDT4TT, respectively.

The Pd impregnation by vapor phase thermal decomposition was carried out following the next steps. After a mechanical mixture of the required amounts of palladium acetyl-acetonate ($Pd(acac)_2$), and the corresponding support to obtain 0.5 wt% loading of $Pd^0$, the method was carried out in two stages. In the first one, the mixture was introduced in a quartz reactor with a porous membrane and it was placed into the system for vapor phase decomposition, at 453 K during 10 min under vacuum conditions. In the second step, the reactor was then placed in a second section of the system in Ar flow (~100 $cm^3$ $min^{-1}$), at 673 K for 10 min more. Finally, these catalyst series were labelled as HDT2-5TD, HDT3-5TD, and HDT4-5TD.

The samples were characterized by X-ray diffraction (XRD) using a D8 Focus Bruker AXS instrument using Cu Kα radiation ($\lambda = 1.5406$ Å, 35 kV, and 25 mA). Infrared spectroscopy (FTIR) was recorded from 4,000 to 400 $cm^{-1}$, using a Magna-IR Nicolet 750 spectrophotometer with a resolution of 4 $cm^{-1}$ (64 scans). $N_2$ sorption experiments at 77 K were carried out on samples previously calcined at 823 K and outgassed ($10^{-4}$ Pa) at 553 K during 8 h with a Micromeritics ASAP 2000 instrument. Specific surface areas were calculated using the BET method and the pore distribution was determined by the BJH method.

The microstructural analysis was studied using a JEOL JSM6701F scanning electron microscope (SEM), working at 5.0 kV. Samples were first recovered with a carbon coat with the aim to increase their conductivity. The instrumental magnification ranged from $5 \times 10^3$ to $10 \times 10^3$. Finally, the morphology of the Pd nanoparticles was studied by high resolution transmission electron microscopy (HRTEM) in a high resolution JEOL 100CX transmission electron microscope, operating at 200 kV. The instrumental magnification ranged at $2 \times 10^5$.

## 6.3 Results and Discussion

X-ray diffraction patterns of as-prepared (a, b, c), and after calcination at 823 K (d, e, f), samples are displayed in Fig. 6.1. HDT samples showed the typical XRD patterns of hydrotalcite-like compounds (JCPDS card 22-0700). The observed reflections are indexed as hexagonal lattices with $R\bar{3}m$ rhombohedral symmetry, as usual for HDT's crystalline structure. It is also noted to the extent that $M^{II}/M^{III}$ molar ratio increases, the intensity of the peaks for the (0 0 3), (1 1 0), and (1 1 3) reflections is better defined. Thus, a better interlayer ordering is obtained when the Mg/Al ratio increases. After calcination, the samples showed similar XRD patterns corresponding to the mixed oxides Mg(Al)O structure, very similar to that one of periclase-like structure MgO (JCPDS card 04-0829).

Additionally, the lattice *a* and *c* parameters and specific surface area of the HDT samples are given in Table 6.1. There is a small shift towards lower $2\theta$ angles as the Mg/Al ratio decreases for the (0 0 3) reflection. This fact can be explained by the effect of the absorbed water degree, depending on the Mg/Al ratio [11, 12]. On the

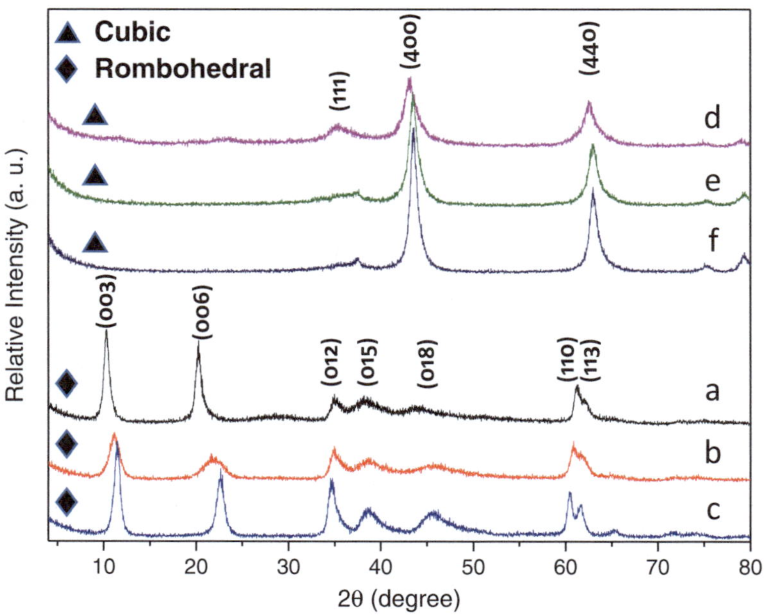


**Fig. 6.1** X-ray diffraction patterns of as-prepared and calcined samples: (**a**) HDT2, (**b**) HDT3, (**c**) HDT4, (**d**) HDT2TT, (**e**) HDT3TT and (**f**) HDT4TT

**Table 6.1** Structural and textural parameters of HDT samples

| Sample | $a$ (Å)[a] | $c$ (Å)[a] | $S_{BET}$ ($m^2\ g^{-1}$)[b] |
|---|---|---|---|
| HDT2 | 3.02 | 25.62 | 162 |
| HDT3 | 3.04 | 23.70 | 151 |
| HDT4 | 3.06 | 23.04 | 149 |

[a]Determined on the layered double hydroxide precursors
[b]Determined on the Mg(Al)O supports

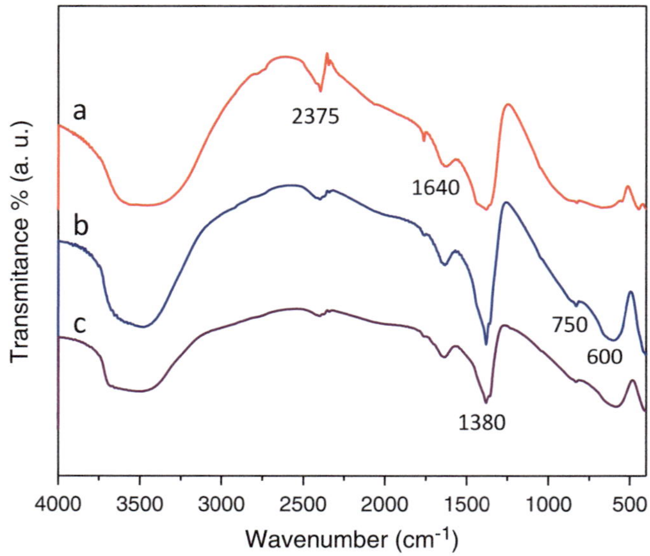


**Fig. 6.2** FT-IR spectra of as-prepared samples: (**a**) HDT2, (**b**) HDT3 and (**c**) HDT4

other hand, lattice $a$ parameter presents a small increase as the Mg/Al ratio does. For the calculation of the lattice $a$ and $c$ parameters, the following equations were used: $a = 2d_{110}$ and $c = 3d_{003}$ [3].

The FT-IR spectra of as-prepared HDT samples are shown in Fig. 6.2. The first broad absorption band around 3,400–3,600 $cm^{-1}$ is assigned to the stretching frequency mode of the O–H functional group in the brucite-like layer as well as for the water placed in the interlayer space. A small absorption band at approximately 2,375 $cm^{-1}$ is also observed. This band can be attributed to the $CO_2$ species adsorbed from the environment since the experiments were not recorded in controlled atmosphere.

The band at approximately 1,640 $cm^{-1}$ is usually assigned to the hydrogen bending vibration of O–H functional groups. It can be also observed a strong absorption band at 1,380 $cm^{-1}$, attributed to the stretching vibration of intercalated $NO_3^-$ and $CO_3^{2-}$ anions, while the band at 750 $cm^{-1}$ is also attributed to the $CO_3^{2-}$ anions. Finally, the bands at 600 and 430 $cm^{-1}$ are ascribed to the lattice vibration modes attributed to M–OH and O–M–O vibrations, where M indicates Mg and Al, coming from the brucite-like layers [13].

**Fig. 6.3** SEM micrographs of HDT2-5TD sample

The morphology of the solids prepared by thermal decomposition impregnation was studied by scanning electron microscopy (SEM). Figure 6.3 shows scanning electron micrographs of the materials with a magnification range of $10 \times 10^3$. In general, all samples present high porosity level. The presence of flake-like conglomerates can be observed in all samples, being the typical morphology of mixed oxides coming from LDH or clay minerals [14–16]. Specifically, sample HDT2-5TD is composed by regular size grains and it is very clear to observe the flake-like conglomerates with dimensions around $0.43 \times 0.37$ μm and 0.06 μm of thickness.

Conventional transmission electron microscopy (TEM) images of a representative sample prepared by thermal decomposition impregnation (HDT4-5TD) revealed the morphology and structure of the Pd nanoparticles on the supports Mg(Al)O. Figure 6.4a shows the conventional bright-field TEM. The sample revealed small average diameter of Pd nanoparticles and high dispersion. They also present the typical structure of mixed oxides coming from LDH. In addition, Fig. 6.4b shows a high resolution transmission electron microscopy (HRTEM) image of a Pd nanoparticle for the corresponding sample. From this image, the *d*-spacings of 0.194 nm and 0.223 nm were obtained; such *d*-spacings correspond to (0 0 2) and (1 1 1) crystalline planes of the Pd structure, according to the fast Fourier transform (FFT) FFT analysis present in Fig. 6.4c, with a value of the lattice parameter $a_0 = 0.389$ nm (JPDF 46-1043). The applied mask in (FFT) Fig. 6.4d confirms the FCC structure belongs to Pd main reflection showed in the sample is the [1 1 0] zone axis.

## 6.4 Conclusions

Pd nanoparticles presented homogeneous distribution and small diameter, both required for the Pd/Mg(Al)O metal-supported catalyst. These parameters are critical attached to the plane of low index in some catalytic reactions depending on the

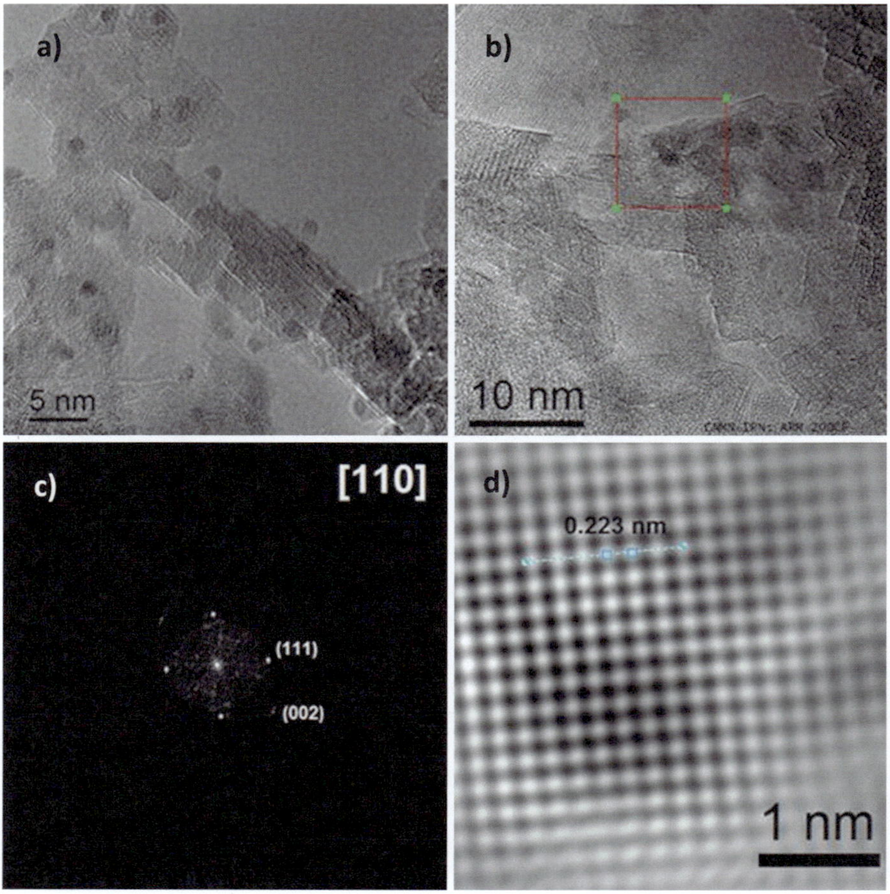


**Fig. 6.4** (**a**) Conventional bright-field TEM image of Pd nanoparticles of the HDT4-5TD sample prepared by vapor phase thermal decomposition impregnation, (**b**) the HRTEM image and (**c**) FTT of region select in image **b** and (**d**) mask apply at image **c**

structure, and they can be controlled by the nucleation and growth processes. Effective catalysts; depending strongly on the synthesis method; in this case, impregnation by vapor phase thermal decomposition were successfully obtained. Thus, the Pd nanoparticles diameters resulted smaller, which makes them excellent catalysts for selective hydrogenation reactions to produce a specific product. Therefore, the size and morphologies of Pd nanostructures could be regulated by changing the experimental parameters, such as the concentrations of metallic precursors, phase of deposition, and the synthesis temperature.

**Acknowledgments** The authors wish to thank the CNMN-IPN, and also wish to thank CONACYT for 61597, INFR-2014-225161 and SIP 20131013 projects for financial support.

## References


1. Xu ZP, Zhang J, Adebajo MO, Zhang H, Zhou C (2011) Catalytic applications of layered double hydroxides and derivatives. Appl Clay Sci 53:139–150
2. Miyata S (1980) Physico-chemical properties of synthetic hydrotalcites in relation to composition. Clays Clay Miner 28:50–56
3. Cavani F, Trifirò F, Vaccari A (1991) Hydrotalcite-type anionic clays: preparation, properties and applications. Catal Today 11:173–301
4. Prinetto F, Manzoli M, Ghiotti G, Martínez-Ortiz MJ, Tichit D, Coq B (2004) Pd/Mg(Al)O catalysts obtained from hydrotalcites: investigation of acid–base properties and nature of Pd phases. J Catal 222:238–249
5. Oliver-Tolentino MA, Guzmán-Vargas A, Manzo-Robledo A, Martínez-Ortiz MJ, Flores-Moreno JL (2011) Modified electrode with hydrotalcite-like materials and their response during electrochemical oxidation of blue 69. Catal Today 166:194–200
6. Lima E, Lasperas M, de Menorval LC, Tichit D, Fajula F (2004) Characterization of basic catalysts by the use of nitromethane as NMR probe molecule and reactant. J Catal 223:28–35
7. Tichit D, Martínez-Ortiz MJ, Francová D, Gérardin C, Coq B, Durand R, Prinetto F, Ghiotti G (2007) Design of nanostructured multifunctional Pd-based catalysts from layered double hydroxides precursors. Appl Catal A 318:170–177
8. Lima E, Guzmán-Vargas A, Méndez-Vivar J, Pfeiffer H, Fraissard J (2008) Fe-ZSM-5 catalysts: preparation in organic media, Fe-particle morphology and NOx reduction activity. Catal Lett 120:244–251
9. Lorret O, Morandi S, Prinetto F, Ghiotti G, Tichit D, Durand R, Coq B (2007) Synthesis and characterization of Pt/Mg(Al)O catalysts obtained from layered double hydroxides by different routes. Micropor Mesopor Mater 103:48–56
10. García-Contreras MA, Fernández-Valverde SM, Vargas-García JR (2010) Pt, PtNi and PtCoNi film electrocatalysts prepared by chemical vapor deposition for the oxygen reduction reaction in 0.5M KOH. J Alloys Compd 504S:S425–S428
11. Pfeiffer H, Martínez-dlCruz L, Lima E, Flores J, Vera MA, Valente JS (2010) Influence of Mg/Al ratio on the thermokinetic rehydration of calcined Mg–Al layered double hydroxides. J Phys Chem C 114:8485–8492
12. Fornasari G, Glfckler R, Livi M, Vaccari A (2005) Role of the Mg/Al atomic ratio in hydrotalcite-based catalysts for NOx storage/reduction. Appl Clay Sci 29:258–266
13. Lima E, Flores J, Cruz AS, Leyva-Gómez G, Krötzsch E (2013) Controlled release of ferulic acid from a hybrid hydrotalcite and its application as an antioxidant for human fibroblasts. Micropor Mesopor Mater 181:1–7
14. Yuan Y, Zhang Y, Shi W (2011) A novel approach for preparing exfoliated UV-cured polymer/LDH nanocomposites via pre-exfoliated organic LDH. Appl Clay Sci 53:608–614
15. Chen X, Mi F, Zhang H, Zhang H (2012) Facile synthesis of a novel magnetic core–shell hierarchical composite submicrospheres $Fe_3O_4$@CuNiAl-LDH under ambient conditions. Mater Lett 69:48–51
16. Del Arco M, Fernández A, Martín C, Rives V (2010) Solubility and release of fenbufen intercalated in Mg, Al and Mg, Al, Fe layered double hydroxides (LDH): the effect of Eudragit® S 100 covering. J Solid State Chem 183:3002–3009

# Chapter 7
# Structural and Optical Characterization of $ZrO_2$ and $Y_2O_3$-$ZrO_2$ Nanopowders

**Nadiia Korsunska, Anton Zhuk, Vasyl Papusha, Oleksandr Kolomys, Yuliya Polishchuk, Yurii Bacherikov, Viktor Strelchuk, Vasyl Kladko, Tetyana Konstantinova, Tetyana Kryshtab, and Larysa Khomenkova**

**Abstract** Structural and optical properties of pure and Y-doped $ZrO_2$ nanopowders with different Y content sintered by co-precipitation of Zr and Y nitrates were investigated. It was observed that the increase of Y content stimulates the transformation of crystalline phase from monoclinic through tetragonal to cubic while the increase of calcinations time leads to the increase of $ZrO_2$ grain sizes. Generally, room temperature photo- and cathodoluminescence spectra showed several bands in blue–orange range. The increase of powder grains results in the enhancement of 2.81-eV photoluminescence caused probably by volume centers. Besides blue–orange emission, additional "red" cathodoluminescence band was observed. Its intensity exceeds essentially the magnitude of other components and increases with cooling. This "red" band was found to be complex and caused by intradefect transition. The excitation mechanism of this band is discussed.

**Keywords** Y-doped $ZrO_2$ • Luminescence • Raman scattering • TEM • Structural and optical characterization

N. Korsunska • A. Zhuk • V. Papusha • O. Kolomys • Y. Polishchuk • Y. Bacherikov
V. Strelchuk • V. Kladko • L. Khomenkova (✉)
V. Lashkaryov ISP of NAS of Ukraine, 45 Pr. Nauky, Kyiv 03028, Ukraine
e-mail: khomen@ukr.net

T. Konstantinova
Donetsk Institute for Physics and Engineering named after O.O. Galkin of the NASU, 72 R. Luxemburg str., Donetsk 83114, Ukraine
e-mail: matscidep@aim.com

T. Kryshtab
Department of Physics, ESFM-IPN, 07738 México D.F., México
e-mail: kryshtab@gmail.com



R. Pérez Campos et al. (eds.), *Materials Characterization*,
DOI 10.1007/978-3-319-15204-2_7

## 7.1 Introduction

Zirconia ($ZrO_2$) has attracted considerable attention because of its mechanical, electric, thermal, and luminescent properties. Emitting $ZrO_2$ allows different applications, for example, for oxygen sensor [1], laser techniques, biological labelling [2], thermoluminescent UV dosimeters, etc. Many different defect-related emission bands can be observed in pure and/or Y-stabilized $ZrO_2$ materials in visible spectral range [3, 4] that offer also an application of zirconia for white light emitting devices [5] and for a visualization of high energy radiation [6].

One of the methods to investigate the availability of $ZrO_2$ to such an application is the study of its cathodoluminescence (CL) which is not well addressed. Only a few works are devoted to CL investigation demonstrating different results [7, 8]. This can be due to different microstructure (type of crystalline lattice, size distribution of $ZrO_2$ grains, and different nature of radiative defects) of $ZrO_2$ materials that depends on the fabrication approach. In particular, freestanding $ZrO_2$ nanocrystals with different sizes, studied in [9], exhibit the lowest CL intensity for the smallest size nanocrystals.

In the present work $ZrO_2$ and $Y_2O_3$-$ZrO_2$ nanopowders with different Y content were investigated by means of Raman scattering, transmission electron microscopy (TEM), photo- (PL) and cathodoluminescence (CL) methods. The effect of fabrication parameters on material crystalline structure and light emission parameters as well as the difference in PL and CL spectra of the same powders were observed. The nature of emission centers and the mechanisms of their excitation are discussed.

## 7.2 Experimental

Undoped and doped nanopowders were synthesized by a co-precipitation technique from using $ZrOCl_2 \cdot nH_2O$ and $Y(NO_3)_3 \cdot nH_2O$ salts. The 25 % $NH_4OH$ water solution was used as precipitant. Sediments were mixed for 1 h at room temperature at pH=9. After this, they were repeatedly washed, filtered with distilled water and then the hydrogel was dried in a microwave furnace and calcined at 700 °C for $t_c = 1$ or 2 h. More details can be found elsewhere [10, 11].

Elemental analysis of powders investigated was performed by means of X-ray fluorescent method using EXPERT 3L W144U setup. Pure $ZrO_2$ ($t_c = 1$ h) and Y-doped $ZrO_2$ (ZYO) powders with 10 at% of Y content ($t_c = 1$ (ZYO-10-1) and 2 h (ZYO-10-2)) as well as with 15 at% of Y content ($t_c = 1$ h (ZYO-15)) were examined. Note that all the samples contained Fe and Zn as impurities.

Micro-Raman scattering spectra were detected using a Horiba Jobin T-64000 Raman spectrometer. A 488-nm line of Ar laser was used as the excitation source. The laser power was kept below 5 mW to prevent laser heating of sample investigated. XRD data were collected using ARL X'TRA powder diffractometer with CuKα wavelength ($\lambda = 0.15418$ nm) and grazing geometry ($\omega \sim 0.5°$). TEM observation was performed with JEOL JEM 200A instrument. For more details, see [12].

Photoluminescence (PL) was excited by 337-nm light of $N_2$-laser. In cathodoluminescence (CL) measurements the samples were excited by scanning electron beam in quasi-continuous regime at accelerating voltage $E=75$ kV and current value ($I$) varied in the range 10–120 μA. PL and CL light was dispersed by grating monochromator and registered by PMT tube. CL was detected at 300 and 77 K, while PL was recorded at 300 K.

## 7.3 Results and Discussion

### 7.3.1 Structural Properties

$ZrO_2$ can exist in the three structural phases: monoclinic (m-$ZrO_2$), tetragonal (t-$ZrO_2$), and cubic (c-$ZrO_2$). At room temperature the m-$ZrO_2$ phase is stable [13], whereas both t-$ZrO_2$ and c-$ZrO_2$ phases can be stabilized by doping with bivalent or trivalent atoms, in particular with Y.

Figure 7.1a represents Raman scattering spectra of samples investigated. Pure $ZrO_2$ powder demonstrates multiple narrow phonon bands which correspond to nine $A_g$ modes at 98, 175, 187, 302, 344, 473, 553, 633, and 706 cm$^{-1}$, and six $B_g$ modes at 219, 330, 378, 499, 533, and 610 cm$^{-1}$ which correspond to $ZrO_2$ monoclinic phase [14].

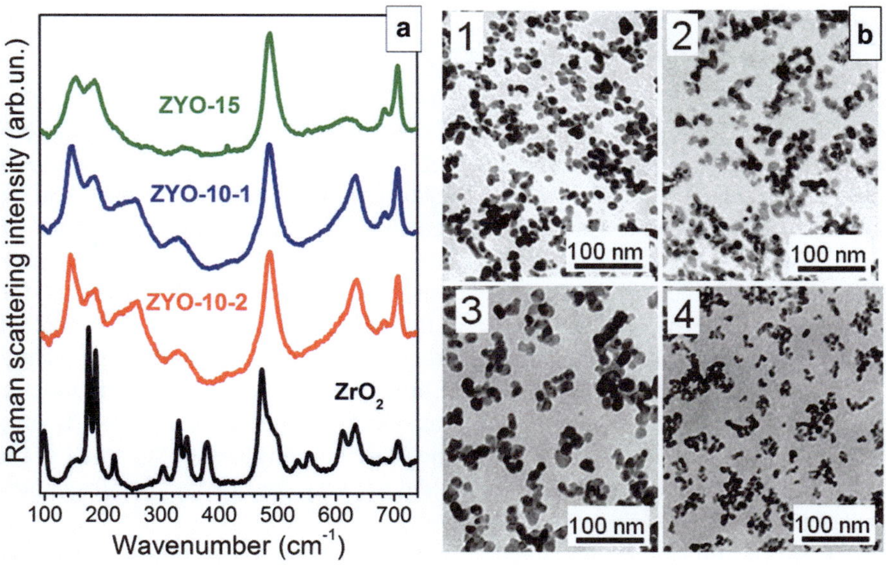


**Fig. 7.1** (**a**) Raman spectra of undoped $ZrO_2$ and yttrium-doped samples (ZYO-10-1, ZYO-10-2 and ZYO-15). Spectral intensities have been normalized to their maximum value. For clarity, the spectra have been shifted in vertical direction. $E_{exc}=2.54$ eV, (**b**) bright-field TEM images of pure ZrO2 (1), ZYO-10-1 (2), ZYO-10-2 (3) and ZYO-15 (4)

The ZYO-10-1 and ZYO-10-2 samples show broad bands peaked at 145, 259, 330, 612 and 636 $cm^{-1}$ attributed to the modes of t-$ZrO_2$, small peaks at 175 and 187 $cm^{-1}$ attributed to $A_g$ modes m-$ZrO_2$ phase. Besides, several broad bands with the peak positions corresponding to m-$ZrO_2$ are also observed in these spectra that can be an evidence of a partial transformation of m-$ZrO_2$ and a coexistence of both t- and m-$ZrO_2$ phases (Fig. 7.1a). The band peaked at 259 $cm^{-1}$ is not observed either for monoclinic or cubic phase [15] and can be exclusively assigned to the $E_g$ mode of t-$ZrO_2$ [16]. The peak's broadening could be due to structural distortion. For ZYO-15 sample the peaks at 155, 279, and 618 $cm^{-1}$ corresponded to c-phase were observed [17]. However, the peak at 485 $cm^{-1}$ is detected also. Therefore, this sample can be considered as superposition of cubic and tetragonal phases with main contribution of cubic one. Observed transformation of crystallite structure is in agreement with XRD data obtained for the same samples and described in detail in [18].

Bright-field TEM images of samples investigated are shown in Fig. 7.1b. It is seen that the powders obtained by co-precipitation technique are single crystals with soft and easily destroyed agglomerates that are important for their successful application. From the analysis of crystal size distribution the mean sizes of nanocrystals were found to be $d=12$ nm for pure $ZrO_2$ sample, $d=11$ nm for ZYO-10-1 sample, and $d=20$ nm for ZYO-10-2 sample as well as $d=10$ nm for ZYO-15 sample. Thus, the increase of Y content stimulates the decrease of nanocrystal sizes, whereas the increase of calcination time results in their enlargement.

### *7.3.2 Light Emitting Properties*

PL spectra of the samples with different Y content are shown in Fig. 7.2. They contain wide structural band in the region of 2.25–3.10 eV (400–550 nm) and the shoulder in the range of 2.06–2.17 eV (570–600 nm). The observed PL band can be approximated by three Gaussian-like components with maxima positions at ~2.81 eV (440 nm), ~2.43 eV (510 nm) and ~2.11 eV (587 mn) (dashed curves). The shape of PL spectra slightly depends on Y content in the samples calcinated for 1 h. The increase of calcination time in the samples with the 10 at% Y content resulted in the increase of the grain size from 11 to 20 nm leads to the high-energy shift of PL maximum due to increase of high-energy component contribution to PL spectra. At the same time the PL components peaked at ~2.11 eV (587 nm) and 2.43 eV (510 nm) are more pronounced in the sample with the small grain size. It should be noted that PL intensity did not vary considerably in all samples investigated.

CL spectra of samples measured at 300 K demonstrate two broad bands with maxima positions at 2.9 eV (425 nm) and 1.74 (700 nm) (blue and red bands) (Fig. 7.3). The intensity of red band exceeds essentially the intensity of blue one. The latter has a long-wavelength tail testifying to its non-elementary. Similar complex

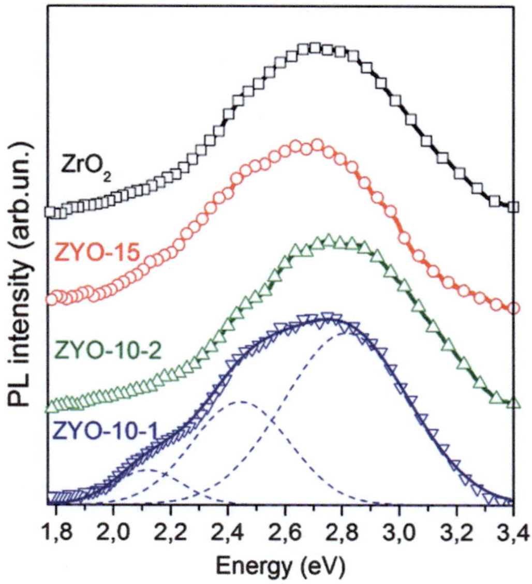


**Fig. 7.2** PL spectra of pure $ZrO_2$ and ZYO powders. *Dashed curves* represent the approximation of PL spectrum of ZYO-10-1 sample by Gaussian-like components

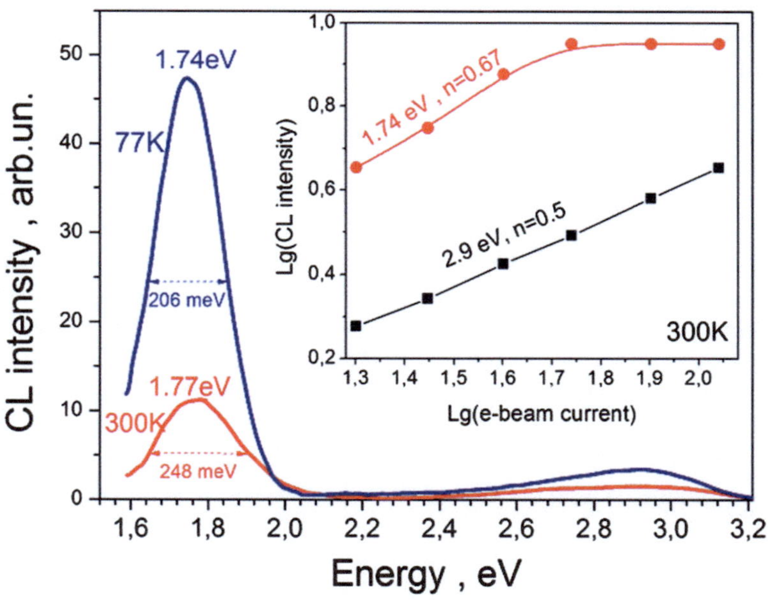


**Fig. 7.3** CL spectra of ZYO powders measured at 300 and 77 K at $E = 75$ kV and $I = 120$ µA. The inset shows the variation of CL intensity for different CL bands (1.74 and 2.9 eV) versus e-beam current at 300 K

band was often observed in photo- [9], X-ray, and cathodoluminescence spectra [7], but the red band, to author's knowledge, was not earlier detected.

The cooling to 77 K enhances the intensity of both bands, but the red band increases mainly. Thus, the both bands show the thermal quenching in the range 77–300 K. Besides, the cooling results in the shift of red band peak position to lower energy side and in the decrease of its full width at half maximum (FWHM) (Fig. 7.3). Therefore we can conclude that it is also not elementary and with temperature decrease its low energy component increases mainly.

The variation of CL band intensities with electron beam current value plotted in log–log scale shows that the dependence for blue band can be approximated by a straight line in the whole range of applied current values, while the red band intensity saturates at $I > 55$ μA (Fig. 7.3 (inset)). The slopes of linear parts of these dependencies are lower than 1.

Under band-to-band excitation, the sub-linearity ($n < 1$) and subsequent saturation can be observed when the filling of emission centers by nonequilibrium carriers approaches to the center concentration [19]. Thus, we can suppose that this case is realized in our CL experiments.

As Fig. 7.2 shows, the PL spectra of all the samples investigated recorded under the excitation below band gap consist of several overlapped PL bands similarly to other data obtained on pure and Y-stabilized $ZrO_2$ [3, 5, 20]. The observed PL bands are usually ascribed to different intrinsic defects such as singly charged oxygen vacancies (F-centers) in the volume of nanocrystals or at their surface [3], as well as $Zr^{3+}$ centers [20] or the different complexes including oxygen vacancies or distortion of crystal lattice due to these vacancies [3, 4].

Invariance of PL spectra shape on the Y content in our samples calcinated during 1 h testifies to the presence of the same defects with nearly the same concentration. At the same time the increase of calcination time resulted in the increase of crystallite sizes leads to the redistribution of PL component intensities. The formation of larger grains is accompanied by the decrease of their surface/volume ratio. This can be the reason of the increase of the contribution of 2.81-eV PL component and decrease of the intensities of 2.11 and 2.43 eV PL components. Thus, the former PL component can be ascribed to the volume defects (for instance, F-like center). This is in an agreement with the conclusion of [5, 21], where similar PL emission was attributed to grain volume defects. At the same time, the 2.11 and 2.43-eV PL components are more pronounced for the samples with smaller grains (Fig. 7.2) that allows attributing them to the grain surface defects. This is in accordance with the data of [7] where the 2.17-eV PL component which is close to that observed in our spectra (Fig. 7.2) was ascribed to defect states at the grain boundaries [3]. At the same time the band at ~2.48 eV (closed to 2.43 eV) was attributed to volume intrinsic defects—$Zr^{3+}$ centers [20].

In CL spectra the blue band peak position is closed to that of high-energy PL component. It means obviously that the same defects are activated under intrinsic and extrinsic excitation light. However, under band-to-band excitation other defect can also be excited resulting in appearance of additional bright emission band in red spectral range.

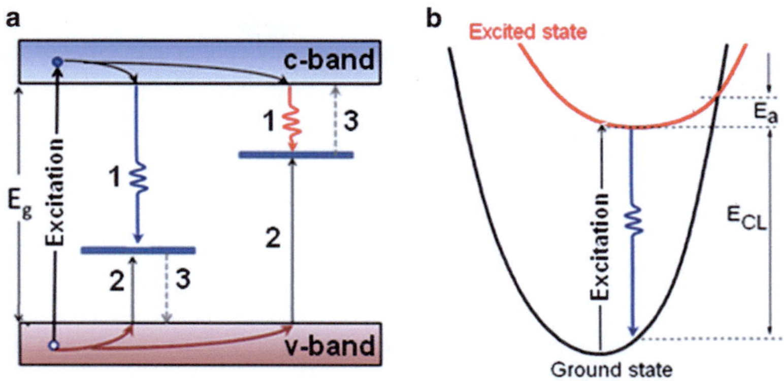


**Fig. 7.4** (**a**) Simplified schema of electron transitions in the case of one-level radiative centers. The labels correspond to capture of free electrons (1) and free holes (2), as well as thermal escape of electrons or holes (3). *Undulating arrows* correspond to radiative transitions. (**b**) The coordination diagram showing the correspondence between excitation, radiative ($E_{CL}$), and nonradiative recombination in two-level center (intradefect transition). The $E_a$ is the activation energy of thermal quenching of emission

As it was mentioned above, both observed CL bands show the thermal quenching above 77 K. Its low temperature range implies that activation energy ($E_a$) of quenching process cannot be high. Based on the comparison with data of [5], we could assume that $E_a \sim 0.2–0.4$ eV.

The thermal quenching can be caused, in general, by two processes (Fig. 7.4). One of them is the escape of electron (hole) into conduction (valence) band and its subsequent capture by nonradiative centers (external quenching) (Fig. 7.4a). Another one is the competition between radiative and nonradiative electron transitions in the same center (intrinsic quenching) (Fig. 7.4b).

The $E_a$ value for external quenching is equal to the distance between the energy level of emission center and nearest permitted band [5]. If the internal quenching takes place, the $E_a$ is the barrier between minimum of potential curve (configuration-coordinate diagram) of excited state and intersection point of potential curves of excited and ground states (Fig. 7.4b). However, irrespectively of quenching mechanism, we can conclude that the recombination center responsible for red emission has, at least, two levels (excited and ground states) because $E_a + E_{CL} < E_g$. Thus, red CL emission is caused by intradefect transition.

Radiative intradefect transition in visible range can be realized in some impurities as well as in the intrinsic defects (oxygen vacancies or $Zr^{3+}$). Currently, the latter emission mechanism is generally accepted for luminescence band(s) in $ZrO_2$. Taking into account the presence of Zn and Fe impurities in our powders, we think that the red CL emission can be caused by intradefect transition either in the impurity or in oxygen vacancies located nearby the impurity.

It is worth to note that this red emission was not detected either under light excitation (whatever excitation light energy below 6.0 eV in our experiment) or under X-ray excitation (with 80 eV energy). Since red emission is the most efficient in CL spectra while the blue band can be stimulated by optical and X-ray excitation, one can conclude that red emitting centers require high-energy electrons to be excited and the main excitation mechanism is impact ionization. Indeed the number of hot electrons in CL experiment is much higher than in X-ray beam.

## 7.4 Conclusions

Luminescence and structural properties of $Y_2O_3$-doped $ZrO_2$ nanopowders sintered by co-precipitation of Zr and Y nitrates with different $Y_2O_3$ content were investigated by Raman scattering, TEM, photo- and cathodoluminescent methods. It was found that at constant calcination temperature, the Y content controls the change of crystalline phase. PL spectra of the samples consist of several overlapped bands in visible spectral range (at ~440, 510 and 560 nm). The contribution of all components to PL spectra does not depend on Y content, but it is affected by calcination time (powder grain sizes). Besides the bands observed in PL spectra, CL ones show additional "red" band which intensity exceeds essentially the magnitude of other CL components. At lower temperatures, preferable enhancement of "red" CL band, its narrowing and peak position shift to the longer wavelengths were found. This behavior testifies to the complexity of "red" CL band which nature and excitation mechanisms are discussed. The dominant contribution of red emission into CL spectra opens some perspectives for its application as a marker for high-energy radiation.

**Acknowledgments** This work is supported by the National Academy of Sciences of Ukraine (project III-4-11).

## References

1. Fidelus JD, Lojkowski W, Millers D, Smits K, Grigorjeva L (2009) Advanced nanocrystalline $ZrO_2$ for optical oxygen sensors. IEEE Sensors 9:1268–1272
2. Jia R, Yang W, Bai Y, Li T (2004) Upconversion photoluminescence of $ZrO_2$:$Er^{3+}$ nanocrystals synthesized by using butadinol as high boiling point solvent. Opt Mater 28:246–249
3. Nakajima H, Mori T (2006) Photoluminescence excitation bands corresponding to defect states due to oxygen vacancies in yttria-stabilized zirconia. J Alloys Compd 408–412:728–731
4. Smits K, Grigorjeva L, Millers D, Sarakovskis A, Grabis J, Lojkowski W (2011) Intrinsic defect related luminescence in $ZrO_2$. J Lumin 131:2058–2062
5. Petrik NG, Tailor DP, Orlando TM (1999) Laser-stimulated luminescence of yttria-stabilized cubic zirconia crystals. J Appl Phys 85:6770–6776
6. Kirm M, Aarik J, Sildos I (2005) Thin films of $HfO_2$ and $ZrO_2$ as potential scintillators. Nucl Instrum Meth Phys Res A 537:251–255
7. Smits K, Millers D, Grigorjeva L, Fidelus JD, Lojkowski W (2007) Comparison of $ZrO_2$:Y nanocrystals and macroscopic single crystal luminescence. J Phys: Conf Ser 93:012035

8. Ken Yueh H, Cox B (2003) Luminescence properties of zirconium oxide films. J Nucl Mater 323:57–67
9. Ramos-Brito F, Garcıa-Hipolito M, Martınez-Martinez R, Martinez-Sanchez E, Falcony C (2004) Preparation and characterization of photoluminescent praseodymium-doped $ZrO_2$ nanostructured powders. J Phys D: Appl Phys 37:L13–L16
10. Konstantinova T, Danilenko I, Glazunova V, Volkova G, Gorban O (2011) Mesoscopic phenomena in oxide nanoparticles systems: processes of growth. J Nanopart Res 13:4015–4023
11. Konstantinova T, Danilenko I, Varyukhin V (2013) Effects of surface and interface in oxide nanoparticle system. Springer Proc Phys 146:135–144
12. Doroshkevich AS, Danilenko IA, Konstantinova TE, Volkova GK, Glazunova VA (2010) Structural evolution of zirconia nanopowders as a coagulation process. Crystallogr Rep 55:863–865
13. Kontoyannis CG, Orkoula M (1994) Quantitative determination of the cubic, tetragonal and monoclinic phases in partially stabilized zirconias by Raman spectroscopy. J Mater Sci 29:5316–5320
14. Anastassakis E, Papanicolaou B, Asher IM (1975) Lattice dynamics and light scattering in hafnia and zirconia. J Phys Chem Solids 36:667–676
15. Nomura K, Mizutani Y, Kawai M, Nakamura Y, Yamamoto O (2000) Aging and Raman scattering study of scandia and yttria doped zirconia. Solid State Ionics 132:235–239
16. Shi L, Tin KC, Wong NB (1999) Thermal stability of zirconia membranes. J Mater Sci 34:3367–3374
17. Gazzoli D, Mattei G, Valigi M (2007) Raman and X-ray investigations of the incorporation of $Ca^{2+}$ and $Cd^{2+}$ in the $ZrO_2$ structure. J Raman Spectrosc 38:824–831
18. Korsunska N, Baran M, Zhuk A, Polischuk Y, Stara T, Kladko V, Bacherikov Y, Venger Z, Konstantinova T, Khomenkova L (2014) Role of paramagnetic defects in light emission processes in Y-doped $ZrO_2$ nanopowder. Mater Res Express 1:045011
19. Lashkarev VE, Lyubchenko AV, Sheinkman MK (1981) Non-equilibrium processes in photoconductors. Naukova dumka, Kyiv
20. Orera VM, Merino RI, Chen Y, Cases R, Alonso PJ (1990) Intrinsic electron and hole defects in stabilized zirconia single crystals. Phys Rev B 42:9782–9789
21. Lin C, Zhang C, Lin J (2007) Phase transformation and photoluminescence properties of nanocrystalline $ZrO_2$ powders prepared via the Pechini-type sol–gel process. J Phys Chem C 111:3300–3307

# Part III
# Characterization of Steels Used in the Oil Industry

This part includes four chapters related to the characterization of steels of common use in the oil and petrochemical industry. The steel characterized includes API X60, X65 and a heat resistant alloy (Cr–Ni). The X60 and X65 steels are used very frequently in the pipelines to transport hydrocarbides. These are steels of high strength and have a good corrosion resistance. The pipeline design must include steel of high resistance to external and internal corrosion and stress corrosion cracking (SCC). External corrosion and SCC phenomena are the main deterioration mechanism of the buried pipelines. The various phenomena that damage the external surface of pipelines are developed in a complex process, mainly due to the heterogeneity of the soils. The three phenomena of deterioration of pipeline steels are treated in the chapters of this part.

Heat-resistant alloys are designed to sustain operation while are exposed to high temperatures. These alloys have widespread uses in petrochemical industry in pyrolysis and furnaces; in addition, these kinds of Cr–Ni alloys can be used in oxidizing, sulfidizing, or carburizing environments. The principal attributes of the alloys are creep strength and corrosion resistance.

The main techniques used in the characterization of these materials are:

- Scanning electron microscopy (SEM).
- X-ray diffraction (XRD).
- Electrochemical impedance spectroscopy (EIS).
- Polarization curves.
- Optical microscopy (OM).
- Mechanical tests.

# Chapter 8
# Improvement of Mechanical Properties of API X-65 Steel by Non-conventional Heat Treatment

**Constantino Natividad Murillo, Rafael García Hernández, Víctor Hugo López Morelos, and Melchor Salazar Martínez**

**Abstract** A non-conventional heat treatment in API X-65 steel samples was carried out by heating the steel up to 1,050 °C and holding for 30 min at this temperature. Subsequently, the samples were water cooled or air cooled. This heat treatment aims to approach the Nb characteristics to the beginning of the solubility at this temperature, for obtaining partial austenitic grain growth. To assess the effect of the heat treatment, tensile testing and hardness measurements were performed as well as metallography. A significant improvement in tensile strength was obtained for samples water cooled, while microhardness was maintained. This behavior is due to the acicular ferrite microstructure obtained. With this microstructure the steel improved its mechanical strength while maintaining its resistance to the stress sulfide cracking (SSC) and therefore will enable reduction of the wall thickness of pipelines.

**Keywords** API X-65 • Non-conventional heat treatment • Mechanical properties • Acicular ferrite microstructure • SSC

## 8.1 Introduction

Recent demands for clean energy and increasing needs for transportation of higher volumes of natural gas and crude oil through high-pressure steel pipelines have led to manufacture of high-strength line pipe steels. The combined accelerated cooling and controlled rolling technology is used to produce high-strength low-alloy (HSLA) steels [1–3]. Crude oil is increasingly showing higher contents of $H_2S$, that

C. Natividad Murillo (✉) • R. García Hernández • V.H. López Morelos
Instituto de Investigaciones Metalúrgicas, UMSNH, Morelia, Michoacán, México
e-mail: consnatividad@yahoo.com.mx; rgarcia@umich.mx; composito@yahoo.com

M. Salazar Martínez
Instituto Mexicano del Petróleo, D.F., México
e-mail: salazarm@imp.mx

 
R. Pérez Campos et al. (eds.), *Materials Characterization*,
DOI 10.1007/978-3-319-15204-2_8

increases the risk of stress sulfide cracking (SSC) in the pipelines, for this reason, the use of API X-65 is limited to sour service. In this context, it is important to develop steels that meet the requirements to resist SSC without sacrificing strength as specified by API standard code [4–6]. The main objective is to increase the operating pressure through thinner pipelines, resulting in an overall reduction in construction and transportation costs. To do this, the best possible combination of strength and toughness levels for pipeline steels is required [7, 8]. The increase in austenisation temperature results in refined grains after continuous cooling to room temperature, that is associated with the appearance of acicular ferrite structure in steel improving its mechanical properties [9]. The main objective of this work is to obtain, via a non-conventional heat treatment, a microstructure of acicular ferrite in order to improve the mechanical strength while maintaining the hardness below the treatment.

## 8.2 Experimental

The material used in the present study was API 5L X-65 pipe with an outside diameter of 609.6 mm and a wall thickness of 14.27 mm. This steel is typically used in the Mexican pipeline system. The chemical composition of this steel is shown in Table 8.1. Specimens of 70 mm × 150 mm were heat-treated at 1,050 °C for 30 min and cooled in water or in air. Heat-treated and as-received plates were machined, in dog bone shape, according to ASTM A370 [10] for tensile testing. Impact test with Charpy V-notch specimens (CVN 10 mm × 10 mm × 55 mm) were carried out at 0 °C according to ASTM E23 [11]. Microhardness measurements were conducted in the as-received steel and in the heat-treated samples. Etching was performed on mirror like surfaces using a solution consisted of 25 g of NaOH and 2 g picric acid diluted in 100 mL of distilled water. The as-received steel was etched for 15 s in 2.5 % nital. Microstructural analysis was carried out by optical (OM) and scanning electron microscopy (SEM).

## 8.3 Results and Discussion

Figure 8.1 shows the microstructure of the API-X65 steel in different conditions: as-received, non-conventional heat-treated water cooled and air cooled. In the as-received material, Fig. 8.1a, it is observed a microstructure characterized by polygonal ferrite grains (F) and perlite bands (P), typical of this type of steel. Figure 8.1b

**Table 8.1** Chemical composition of X-65 steel, wt.%

| Steel | C | Mn | Si | P | S | Al | Nb | Cu | Cr | Ni | V | Ti |
|---|---|---|---|---|---|---|---|---|---|---|---|---|
| X-65 | 0.04 | 1.48 | 0.25 | 0.012 | 0.002 | 0.041 | 0.047 | 0.09 | 0.02 | 0.5 | 0.069 | 0.017 |

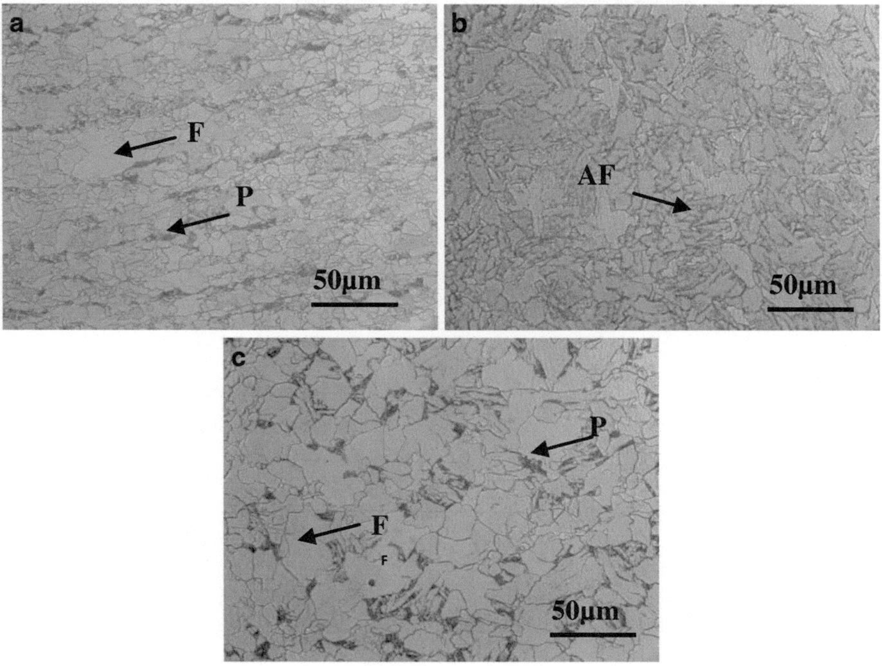


**Fig. 8.1** Optical images of the microstructure of API X-65 steel (**a**) as-received, (**b**) heat-treated water cooled and (**c**) heat-treated air cooled: F-ferrite (*gray phase*); P-perlite (*dark phase*) and AF-acicular ferrite

shows a microstructure composed of fine acicular ferrite (AF) as a result of the non-conventional heat treatment when the sample is cooled rapidly with water. This type of microstructure is preferred to reduce the susceptibility to SSC, as reported by Rakhshkhorshid and Hashemi [12], Min et al. [13], and Danaf et al. [14]. Figure 8.1c reveals that when the steel is cooled more slowly, the microstructure obtained corresponds to the one for a normalizing heat treatment, that is, polygonal ferrite and perlite areas with larger grain size than the as-received material.

Figure 8.2 shows the stress–strain curves obtained for API X-65 with and without heat treatment. The tensile properties are summarized in Table 8.2 along with impact results. Unfortunately, failure of the as-received steel and heat-treated air cooled specimens occurred beyond the limits of the clip-on extensometer and the complete behavior of the test was not followed.

The tensile tests obtained show that heat-treating with cold water leads to a higher tensile strength as is reported by Sha [15], Ray et al. [16], and NACE Standard Test Method TM0175 [17]. Interestingly, the increment in strength does not mean a sacrifice in ductility due to the acicular ferrite microstructure obtained with the non-conventional heat treatment. API specification 5L for pipe line indicates a minimum of 22 % in elongation. According to the literature, for the impact test the minimum impact energy is 73 J [4], in the present work it was obtained 164 J of energy

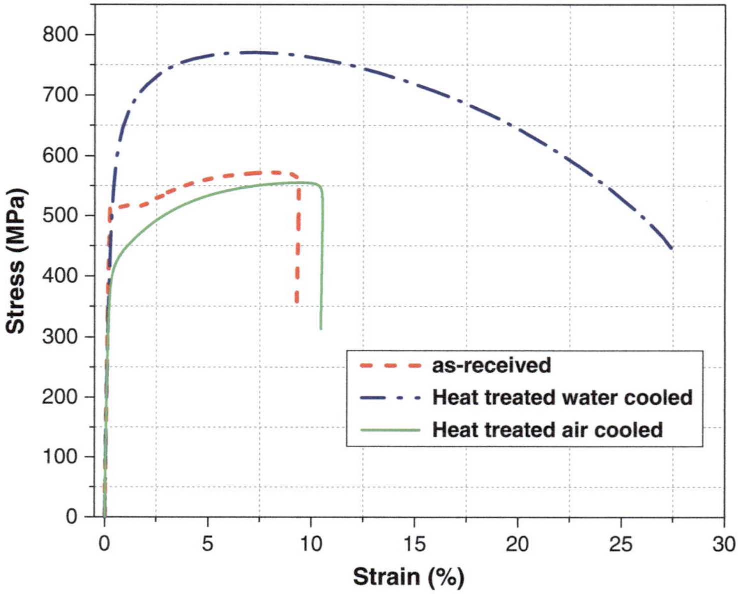


**Fig. 8.2** Stress–strain curves for API X-65 steel

**Table 8.2** Summary of mechanical properties obtained for API X-65 steel with and without heat treatment

| Steel (X-65) | UTS (MPa) | YS (MPa) | $\sigma_{RUPTURE}$ (MPa) | EL (%) | EA (J) |
|---|---|---|---|---|---|
| As-received | 572 | 511 | 350 | – | 228 |
| Heat-treated water-cooled | **771** | **566** | **450** | **27.5** | **164** |
| Heat-treated air-cooled | 555 | 407 | 325 | – | 270 |

*UTS* ultimate tensile strength, *YS* yield strength, *σRUPTURE* stress at rupture, *EL* elongation, *EA* energy absorbed

absorbed. In the fracture surfaces analyzed by SEM, it is evident that both specimens exhibited a kind of ductile fracture, since they showed micro plastic deformation, nucleation and coalescence of microvoids around of some nonmetallic inclusions (Fig. 8.3a–c). These inclusions were identified by energy dispersive X-ray spectroscopy (EDX), attached to the SEM, as iron–manganese sulfides. Figure 8.3d shows the EDX spectra corresponding to these inclusions. The inclusions are mainly composed by Fe, Mn, S, Mg, and Ca. These elements can form the following composites: (Fe, Mn)S, MnS and globular inclusion or Fe–Mn–S as reported in the literature [18].

The heat treatment applied in this study was referred to as non-conventional because the holding temperature is higher than the usual quenching or normalizing temperature which is in the range of 850–900 °C. In this instance, the temperature is raised to 1,050 °C in order to take advantage of the onset of the solubility of

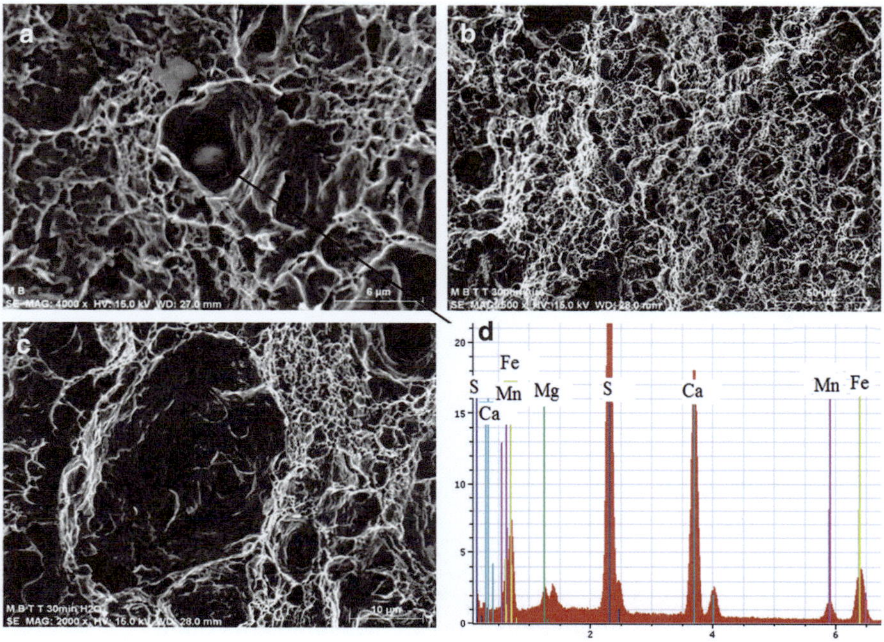


**Fig. 8.3** SEM micrographs after being fractured by tensile tests, the micrographs (**a**) and (**c**) show a globular inclusion and (**b**) shows the typical ductile fracture and (**d**) X-ray compositional spectrum of inclusions shown in (**a**) and (**c**)

carbides or carbonitrides of niobium. At this specific point, partial grain growth of austenite occurs, with the formation of small grain surrounding large grains and the rapid cooling induced by water yields a microstructure of acicular ferrite. This type of microstructure may also be obtained at higher temperatures or at 1,050 °C or below with larger holding times at temperature but excessive grain growth decreases the mechanical properties in every case.

The typical microstructure that dictates the nominal mechanical properties of API X-65 steel after the thermomechanical forming process, Fig. 8.1a, is composed of ferritic grains with scattered perlite. The microstructure of fine acicular ferrite, Fig. 8.1b, obtained with the non-conventional heat treatment leads to an increase of approximately 200 MPa in strength as compared to the as-received material. The heat-treated material with slow cooling, Fig. 8.1c, gives rise to a microstructure corresponding to a normalized heat treatment carried out at elevated temperature, with large ferritic grains and hence lower mechanical properties in comparison with both as-received steel and non-conventional heat-treated steel water cooled.

The limits of hardness recommended for avoiding SSC are in the range of 20–30 HRC (240–250 Hv), but the 22 HRC and its equivalent in other hardness scales is the more common reported by the standard [19]. Although the limits of hardness are not always an effective guide for the selection of a material, these limits are the

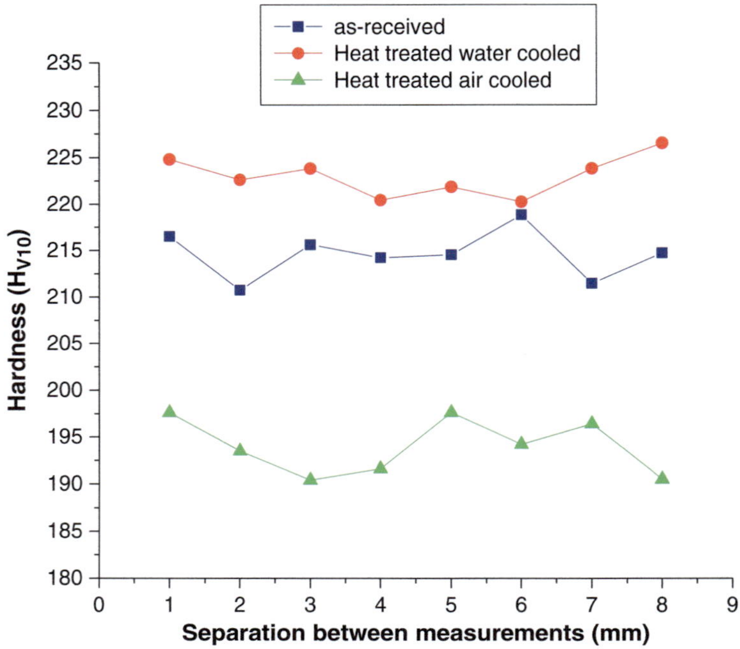


**Fig. 8.4** Hardness measurements in as-received and heat-treated API X-65 steel

more common requirements for the line pipes in sour service. The results of the hardness measurements obtained are presented in Fig. 8.4. Although the hardness of the specimen heat-treated water cooled is increased in comparison with the as-received steel, its average value remains below the critical limit for the susceptibility to the SSC.

## 8.4 Conclusions

Heat-treating an API 5L X-65 pipe steel to 1,050 °C for 30 min following water cooling yields an acicular ferrite microstructure with a mechanical strength 200 MPa larger than the as-received steel. Although hardness also increases, the average value is below the recommended to avoid susceptibility to SSC. The results obtained with the non-conventional heat treatment yield similar properties to an API 5L X-80 steel; however, the high hardness of this steel prevents its use with sour hydrocarbons. The findings of this study imply that it is possible to reduce the wall thickness of pipelines to transport crude oil while maintaining its resistance to SSC.

**Acknowledgments** The authors would like to thank CONACYT for financial support and the IIM-UMSNH for technical support.

## References


1. Hwang B, Kim Y, Lee S, Kim N, Ahn S (2005) Correlation of microstructure and fracture properties of API X70 pipeline steels. Metall Mater Trans A 36:725–739
2. Baek J, Kim Y, Kim C, Kim W, Seok C (2010) Effects of pre-strain on the mechanical properties of API 5L X65 pipe. Mater Sci Eng A 527:1473–1479
3. Ishikawa N, Endo S, Kondo J (2006) High performance UOE linepipes. JFE Technical Report No. 7, pp 19–24
4. American Petroleum Institute (2013) Specification for line pipe, API Specification 5L, 45th Edition, USA
5. Kyeong D, Gyo H, Young K, Jim S (2013) Effect of heat treatment on hydrogen diffusion and hydrogen induced cracking behavior of process pipe steel in sour environment. Proceedings of the 23th international offshore and polar engineering, USA, pp 30–36
6. Han D, Kim S, Kim W, Kang K, Kim K (2012) Effect of pipe forming on hydrogen permeation behavior of linepipe steel in sour environment. Proceeding of the 22nd Int. Offshore Conference, pp 8–15
7. Hwang B, Kim Y, Lee S, Kim Y, Kim N, Yoo J (2001) Effective grain size and Charpy impact properties of high-toughness X70 pipeline steels. Metall Mater Trans A 36:2107–2114
8. Harksen S, Lischewski I, Shneider A, Schmidt T, Mahn D (2013) A modern methodology of materials development for high toughness seamless line pipe products for offshore applications. Proceedings of the 23th international offshore and polar engineering, USA, pp 24–36
9. Zhang D, Qiao Z, Liu Y (2013) Effect of austenisation temperature on phase transformation in low carbon microalloyed pipeline steel. Mater Res Innov 17:S200–S204
10. ASTM A 370 (2014) Standard test methods and definitions for mechanical testing of steel products, Edition 14, May 2014, ASTM International, United States, Pages: 50.
11. ASTM E 23 (2013) Standard test methods for notched bar impact testing of metallic materials, Edition 12C, November 2012, ASTM International, United States, Pages: 25.
12. Rakhshkhorshid M, Hashemi S (2013) Experimental study of hot deformation behavior in API X65 steel. Mater Sci Eng A 573:37–44
13. Min Z, Zhiyong L, Culwel D (2013) Stress corrosion cracking behavior and mechanism of X65 and X80 pipeline steels in high pH solution. Acta Metall Sin 49:1590–1596
14. Danaf E, Muneer B, Abdulhakim A (2013) Mechanical microstructure and texture characterization of API X65 steel. Mater Design 47:529–538
15. Hashemi S (2011) Strength-hardness statistical correlation in API X65 steel. Mater Sci Eng A 528:1648–1655
16. Hashemi S, Mohammadyani D (2012) Characterization of weldment hardness, impact energy and microstructure in API X65 steel. Int J Press Vessels Pip 98:8–15
17. Sha W (2013) Steels: from Mater. Sci. to Structural Eng. Chapter 2 (High-Strength Low-Alloy Steel). Springer, London
18. Ray A, Paul S, Jha S (1995) Effect of inclusions and microstructural characteristics on the mechanical properties and fracture behavior of HSLA steel. J Mater Eng Perform 4:679–688, Springer
19. NACE Standard Test Method TM0175 (2009) Sulfide stress cracking resistant metallic materials for oilfield equipment. NACE, Houston

# Chapter 9
# Aging of Cast Heat Resisting Alloys 35Cr–45Ni–0.1C (MORE40X) and 40Cr–45Ni–0.2C (UCX)

**Ireri Aydée Sustaita Torres, Sergio Haro Rodríguez, and Rafael Colás Ortiz**

**Abstract** Heat resisting cast alloys are designed to sustain operation while are exposed to temperatures greater than 650 °C. These alloys have widespread uses in petrochemical industry in pyrolysis and reformer furnaces, etc.; in addition, they can be hold into oxidizing, sulfidizing, or carburizing environments. The principal attributes of the alloys are creep strength and corrosion resistance. The purpose of this study was to explore the effects of chromium contents in two centrifugal cast pipes of Ni-base heat resistant alloys, one MORE40X 0.1C and the other UCX 0.2C during aging. The behavior of this alloys during aging was examined by optical and scanning electron microscopy in samples aged at 750 °C up to 1,000 h. The microstructural evolution was analyzed on selected samples and images using secondary and backscattered electron detectors and with X-ray energy dispersive spectroscopy. The main microstructural changes in primary and secondary carbides that occurred during aging were described and related to mechanical properties in two alloys. The microstructure as cast materials shows a primary carbides network, in austenitic matrix. The aging times produced different changes in the alloys, which dependent of the composition, morphology, distribution and in the transformation of carbides inside them. It was found that aging promoted the increment in tensile strength and the reduction in ductility. However the mechanical properties of UCX (aged at 750 °C up to 1,000 h) are better with respect to MORE40X (as cast).

**Keywords** Heat resisting cast alloys • Strength • Corrosion resistance • Optical • Scanning electron microscopy (SEM)

---

I. A. Sustaita Torres (✉) • S. Haro Rodríguez
Unidad Académica de Ingeniería, Universidad Autónoma de Zacatecas, 98000 Zacatecas, México
e-mail: ireri.sustaita@gmail.com; haros907@hotmail.com

R. Colás Ortiz
Facultad de Ingeniería Mecánica y Eléctrica, Universidad Autónoma de Nuevo León, 66450 San Nicolás de los Garza, México

Centro de Innovación, Investigación y Desarrollo en Ingeniería y Tecnología, Universidad Autónoma de Nuevo León, 66630 Apodaca, México
e-mail: colas.rafael@gmail.com; rcolas@mail.uanl.mx



R. Pérez Campos et al. (eds.), *Materials Characterization*,
DOI 10.1007/978-3-319-15204-2_9

## 9.1 Introduction

Heat resistant alloys produced by centrifugal casting have great development to be used in applications with higher pressures and temperatures. These alloys have been produced by centrifugal casting for more than 50 years and are used in conditions that involve high pressures and temperatures. Early work involve alloys of the HK type (25Cr-20Ni), being followed later on by the HP type (25Cr-35Ni). Demands for the reduction in fuel and emissions impose harsher operating conditions in many industrial sectors, and so the development of a new generation of heat resistant alloys with contents of 35% Cr and 45% Ni, as well as and many other elements [1–5]. Heat resistant alloys are materials designed to withstand oxidizing and corrosive environments in sustained operation at temperatures greater than 650 °C; most of these materials are cast from melts rich in iron, chromium, and nickel. Elements such as niobium, titanium, vanadium, and zirconium are commonly added to impart higher creep resistance, as they form stable precipitates at the operating temperatures. On the other hand, silicon and manganese are added to increase resistance to different degrading atmospheres. Due to their exceptional resistance to carburization and oxidation up to 1,149 °C, heat resistant alloys are used as tubes, flanges, cones, harps, headers and coils in furnaces, transfer lines, heaters, and other appliances, for direct reduction of iron ore plants, petrochemical, chemical, and commercial heat treating industries.

Alloys with high contents of chromium and nickel have an as cast microstructure of austenitic dendrites and a network of primary carbides. Exposure at high service temperatures may promote dissolution of primary carbides. The precipitated carbides coarsen at prolonged exposure times, affecting the mechanical properties of the alloy [1–5]. The aim of this work was to explore the effects of chromium contents in two centrifugal cast pipes of Ni-base heat resistant alloys, one MORE40X 0.1C and the other UCX 0.2C during aging for periods up to 1,000 h at 750 °C.

## 9.2 Experimental

The samples were cut out from heat resistant alloys 35Cr–45Ni–0.1C (MORE40X) and 40Cr–45Ni–0.2C (UCX), and were held at 750 °C for up to 1,000 h on air in an electric resistance furnace, see Table 9.1 for composition information. The specimens were prepared for their metallographic examination following standard polishing procedures and were etched with an electrolytic solution of 10 g oxalic acid in 100 mL water with a stainless steel cathode at 26 °C, a potential of 6 V was applied for 3–5 s.

The microstructure of specimens in the as cast conditions and aged for 500 and 1,000 h was examined in an inverted optical microscope (OM). A scanning electron

**Table 9.1** Chemical composition of the alloys studied

| Alloy | C | Si | Mn | Cr | Ni | P | S | Nb | Ti |
|---|---|---|---|---|---|---|---|---|---|
| UCX | 0.2 | 2.22 | 1.25 | 39.59 | 45.4 | 0.001 | 0.001 | 0.07 | 0.22 |
| MORE40X | 0.15 | 1.53 | 1.03 | 34.02 | 48.09 | 0.02 | 0.01 | 1.31 | 0.08 |

microscope (SEM) was used to evaluate the microstructure of selected samples using secondary (SE) and backscattered electron (BE) detectors; X-ray analyses of selected areas were also obtained (EDX).

Standard tensile specimens according to ASTM E8 [6] were machined and tested at room temperature. Fracture surface of tensile specimens in the as cast and aged condition were observed by SEM to evaluate the fracture mode.

## 9.3 Results and Discussion

The optical microscopy of as cast samples show that the microstructure of the alloys is made of an austenite matrix and a network of primary carbides present in the interdendritic areas, Fig. 9.1a. Scanning electron microscopy examination, using

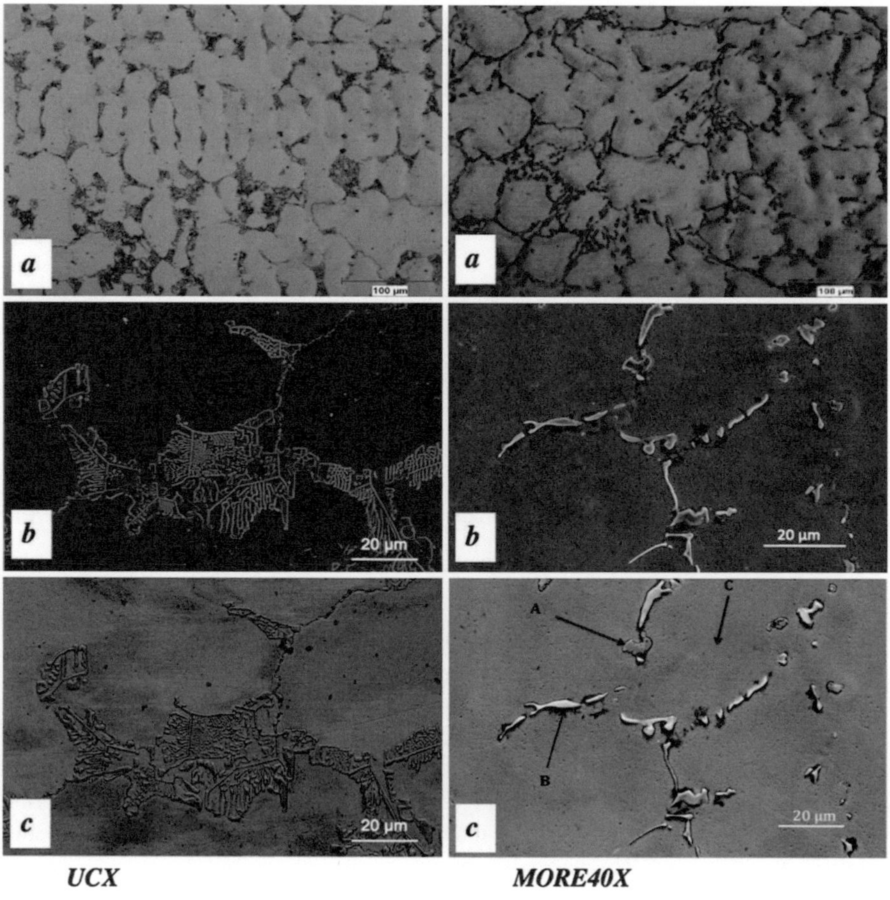


**Fig. 9.1** Microstructure of originals as cast alloys. (**a**) Optical micrograph 200×, and SEM 1,000× in, (**b**) Secondary (SE) and, (**c**) Backscattered (BE) imagining modes

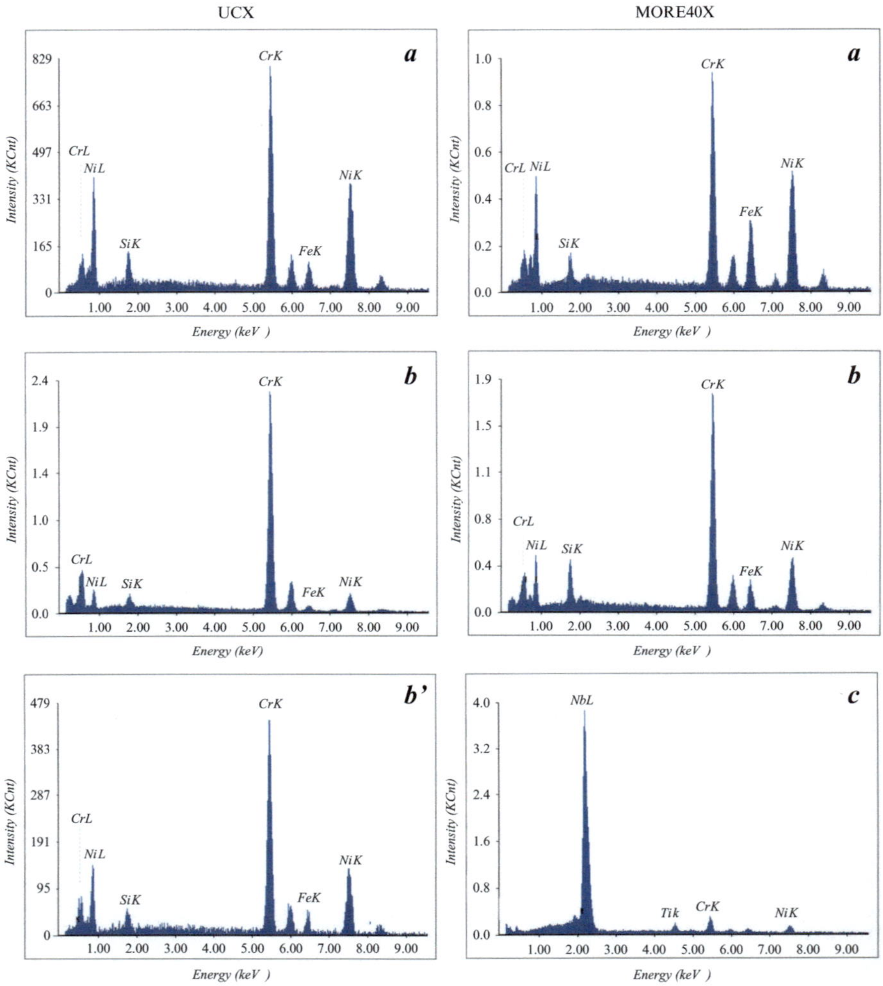


**Fig. 9.2** EDX spectra from the (**a**) matrix, obtained by selected area analysis in (**b**) and (**b′**) dark carbides and (**c**) bright carbides; of originals as cast alloys

secondary electron (SE) Fig. 9.1b, does not distinguish between the different types of carbides present, but using backscattered electron (BE) detectors is possible distinguish between the two different types of carbides present in the alloy MORE40X, and for alloy UCX only one type of primary carbides, Fig. 9.1c.

The EDX analysis carried out on the matrix and on either type of primary carbide of the alloys UCX and MORE40X (Fig. 9.2) shows that the dark carbides are rich in chromium (Fig. 9.2b, b′), whereas the bright particles contain Nb (Fig. 9.2c) and confirm that the alloy UCX has only one type of primary carbides.

The aging promoted changes in the microstructure of the alloys in Fig. 9.3 show the optical and SEM images from a sample held for 1,000 h at 750 °C. The most

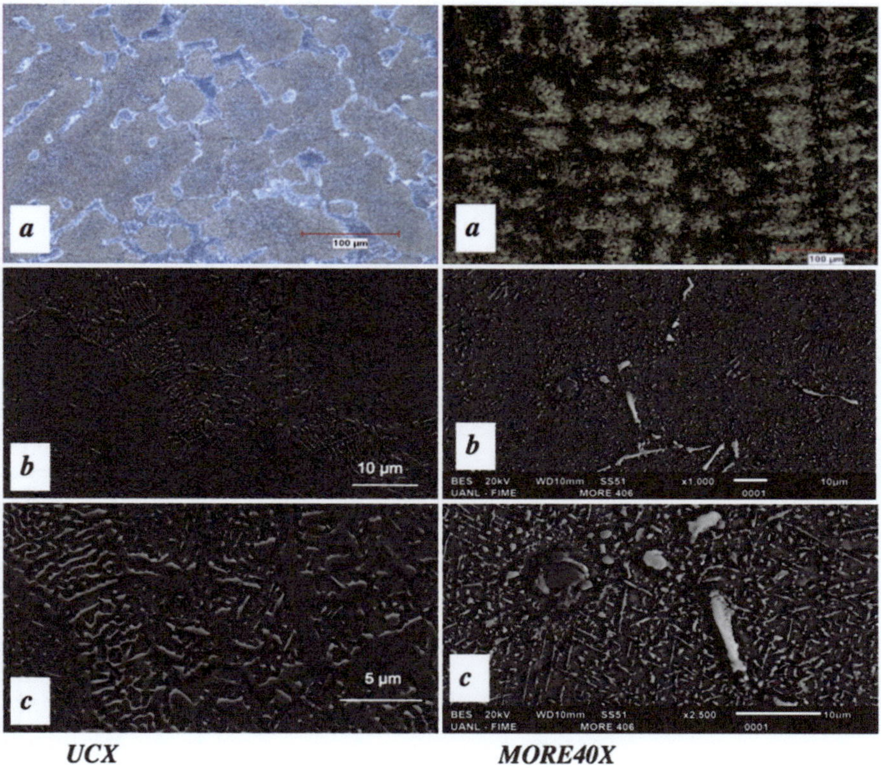


**Fig. 9.3** Images obtained from samples held for 1,000 h at 750 °C (**a**) Optical micrograph 200×, and SEM in, (**b**) backscattered (BE) imagining modes 1,000×, (**c**) backscattered (BE) imagining modes 2,500×

remarkable features in the aged samples are the precipitation of small needle-shape and globular secondary particles. It is observed that the secondary carbides have the tendency to grow along well-defined directions, Fig. 9.3c (UCX and MORE40X); the angles between different needle-shape particles were measured, and it was found that particles were growing following directions that were inclined either 60° or 90° to each other, indicating that the particles grow following ⟨110⟩ directions (it can be deduced by crystallographic means that any two different [110] directions are inclined either 60° or 90° with respect to each other). One characteristic of the cubic system is that planes and directions normal to each other have the same index, therefore, it can be deduced that the growing particles are of the $Cr_{23}C_6$ type, as these carbides nucleate and grow on {110} planes [7]. In accordance with the EDX analysis (see Table 9.2) of phases in alloy aged at 750 °C, 1,000 h, the secondary carbides contain Ni and Cr. For the alloy MORE40X the secondary carbides contents of carbon is around 20 % which corresponding to particles of $M_{23}C_6$ type. However the contents of carbon for the secondary carbides are higher for the alloy UCX, because to the different chemical composition, its no possible identifies the particles as $M_{23}C_6$ type.

**Table 9.2** EDX Analysis of secondary carbides in alloys aged at 750 °C, 1,000 h

| Alloy | C | Fe | Ni | Cr | Si |
|---|---|---|---|---|---|
| UCX Globular secondary carbides | 58.74 | 2.62 | 16.51 | 14.78 | 1.45 |
| UCX Needle-shape secondary carbides | 59.11 | 2.57 | 15.49 | 14.58 | 1.96 |
| MORE40X Globular secondary carbides | 23.25 | 6.3 | 11.06 | 37.61 | 1.6 |
| MORE40X Needle-shape secondary carbides | 28.28 | 6.47 | 13.67 | 36.28 | 4.8 |

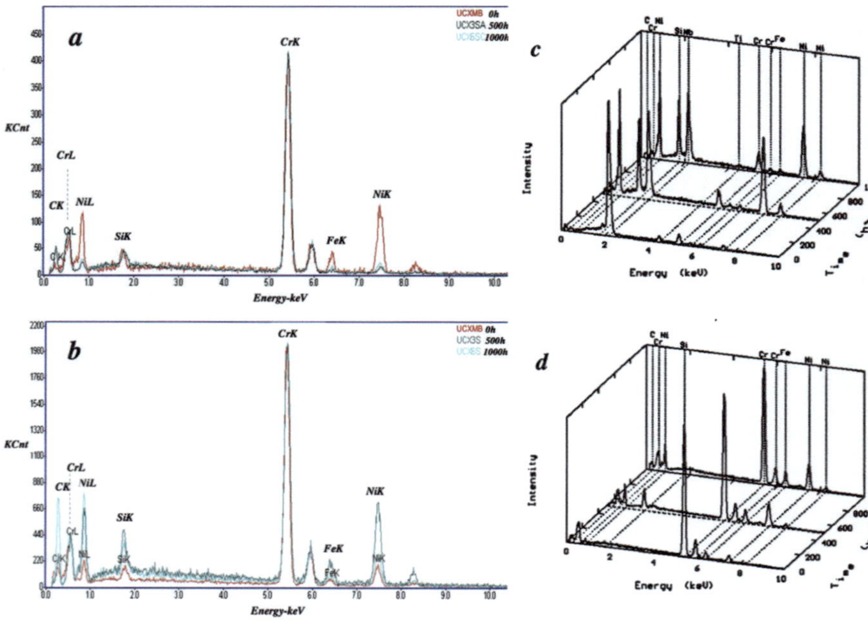


**Fig. 9.4** X-ray spectra obtained from point analysis of (**a**) dark gray primary carbides, (**b**) choral shape primary carbides, in UCX; and (**c**) bright primary carbides, (**d**) dark primary carbides, in MORE40X; materials aged for 1,000 h at 750 °C

Figure 9.4a (dark gray primary carbides) and Fig. 9.4b (choral shape primary carbides) show the representative EDX spectra for primary carbides as a function of aging time at 750 °C (UCX); in these spectra, one can see that the amount of nickel of the particles diminishes a bit as a function of time. The aging, for 1000 h at 750ºC, produces changes in the alloy MORE40X in the bright primary carbides, because the amounts of Ni and Si are increased; whereas the dark primary carbides remain without noticeable change (Fig. 9.4c, d).

The fracture surface of the tensile samples was studied by SEM which permits to identify the mode of fracture. Figure 9.5 shows the images in secondary mode (SE) of the samples in their as cast and aged (1,000 h at 750 °C) conditions. The image

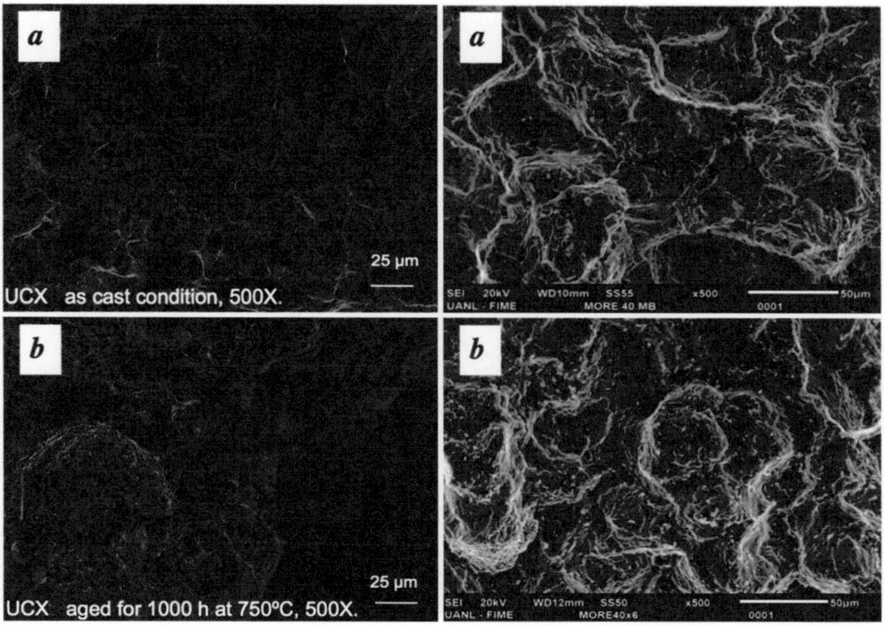


**Fig. 9.5** (**a**) Fracture surfaces of samples tested in as cast condition, 500×. (**b**) Fracture surfaces of samples tested in aged for 1,000 h at 750 °C, 500×

of the surface of the as cast specimens, Fig. 9.5a, reveals a mixed mode of fracture. Most of the surface is of the fragile type (of faceted brittle appearance) but a series of small dimples (indicative that plastic deformation took place prior to fracture) are also appreciated. The image of the facture surface of the aged specimen, Fig. 9.5b, reveals a well-faceted brittle appearance, without any plastic deformation prior to fracture, an indication that the austenitic matrix has lost its capacity to deform plastically. These changes are less drastic in alloy UCX.

The main effect of aging and the change in microstructure is the reduction in ductility and the increase in tensile strength for both alloys (UCX and MORE40X), see Fig. 9.6. These results coincide with those obtained in similar materials [8, 9]. The loss in ductility is important when these types of alloys have to be repaired by welding after service, as the low value of ductility will not allow the repair without an adequate annealing heat treatment [10, 11]. It has been claimed that heat resistant cast steels are very sensitive to weld cracking, as cracks are formed in the brittle zones created by primary carbides [12]. It is observed that the mechanical properties of UCX (aged at 750 °C up to 1,000 h) are better with respect to MORE40X (as cast).

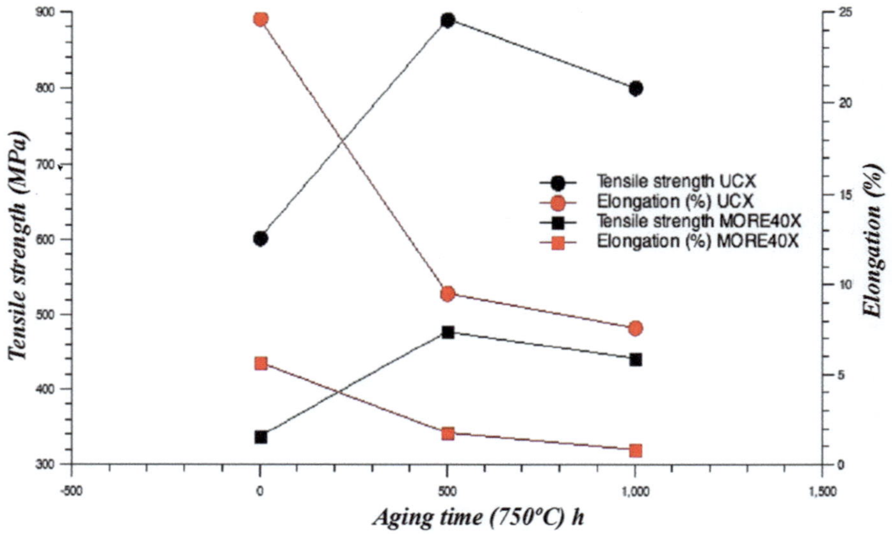


**Fig. 9.6** Changes in mechanical properties as a result of aging

## 9.4 Conclusions

The alloys 35Cr–45Ni–0.1C (MORE40X) and 40Cr–45Ni–0.2C (UCX) are a complex material, containing different phases with high temperature transformations. The analysis shows in the cast state the alloys comprise austenitic matrix and a network of primary precipitates present in the interdendritic areas.

In the alloy 40Cr–45Ni–0.2C (UCX) after aging secondary precipitates were found, whereas the primary phases show no significant changes in their composition. On other hand, in the alloy 35Cr–45Ni–0.1C (MORE40X) after aging secondary precipitates were found, whereas the primary phases show changes in their composition. Analysis by SEM and EDX indicates that the dark primary carbides, rich in Cr, were not affected significantly, whereas the bright primary carbides, rich in Nb, showed significant changes in their composition (the contents of Ni and Si are increased).

The aging promoted the increment in tensile strength and the reduction in ductility. However the mechanical properties of UCX are better with respect to MORE40X.

**Acknowledgments** This work was supported partially by Universidad Autónoma de Zacatecas and Consejo Zacatecano de Ciencia y Tecnología (COZCyT).

## References


1. Garbiak M, Jasinski W, Piekarski B (2011) Materials for reformer furnace tubes. History of evolution. Arch Foundry Eng 11:47–52
2. Hou WT, Honeycombe RWK (1985) Structure of centrifugally cast austenitic stainless steels. HK 40 as cast and after creep between 750 and 1000°C. Mater Sci Technol 1:385–389
3. Kirchheiner R, Woelpert P (2001) Niobium in centrifugally cast tubes for petrochemical applications. International symposium, Niobium; science & technology, pp 1041–1054
4. Zhu SJ, Zhao J, Wang FG (1990) Creep Crack Growth of HK40 Steel: Microstructural Effects. Metallurgical Transsactions A 21:2237–2241
5. Barbabela GD, Almeida LH, Silveira TL, Le May I (1991) Role of Nb in modifying the microstructure of heat-resistant cast HP steel. Materials Characterization 26:193–197
6. ASTM Standard E8/E8M-08 (2008) Standard methods for tension testing of metallic materials. ASTM Int, West Conshohocken. www.astm.org. doi:10.1520/E0008_E0008M-09
7. Martin JW (1980) Micromechanisms in particle-hardened alloys. Cambridge University Press, Cambridge
8. Molina JO (1992) PhD Thesis. Facultad de Ingeniería Mecánica y Eléctrica, Universidad Autónoma de Nuevo León, México
9. Rodríguez J, Haro S, Velasco A, Colás R (2004) Aging of cast Ni-base heat resisting alloy. Int J Cast Met Res 17:188–192
10. Ebert HW (1976) Fabrication of HK-40 in the field. Weld J 55:939–945
11. Haro S, López D, Velasco A, Viramontes R (2000) Microstructural factors that determine the weldability of a high Cr-high Si HK 40 alloy. Mater Chem Phys 66:90–96
12. Duchosal A, Deschaux-Beaume F, Bordreuil C, Fras G (2008) Method for predicting risk of cracking during weld repair of heat resistant cast steels. Materials Characterization 13:126–135

# Chapter 10
# SCC of X-65 Weldment Assessment in Diluted $NaHCO_3$ Solutions with Chloride and Sulfate Ions

**M.A. Espinosa-Medina, G. Carbajal-De la Torre, C. Ángeles-Chavez, and J.G. González-Rodriguez**

**Abstract** In the petroleum Mexican industry, most transportation of hydrocarbons is carried through underground pipelines, which are protected by coating and cathodic protection; however, pipelines could be susceptible to the stress corrosion cracking (SCC) and hydrogen embrittlement damage. In this sense, SCC assessment of API X-65 pipeline weldments in sodium bicarbonate solutions at 50 °C using slow strain rate tests (SSRT) was realized. SSR tests were complemented with hydrogen permeation and electrochemical polarization evaluations. Results showed that corrosion rate, $i_{corr}$, was highest in both 0.01 M $NaHCO_3$ and 0.01 M $NaHCO_3$ with additions of $SO_4^{2-}$ ions solutions. The SCC susceptibility, in accord with the reduction in area percentage (RA%), was higher within the 0.01 M $NaHCO_3$ and $NaHCO_3$ with $Cl^- + SO_4^{2-}$ ions addition solutions. The amount of hydrogen uptake for the weldment was also higher in the 0.01 M $NaHCO_3$ solution with and without $Cl^- + SO_4^{2-}$ ions. The SCC susceptibility of X-65 weldments in diluted $NaHCO_3$ solutions was associated with the anodic dissolution, assisted by hydrogen entrapping within the weldment microstructure defects, as hydrogen embrittlement damage. Results were complemented by Scanning Electronic Microscopy and Energy of Dispersion of X-Ray Spectroscopy characterization.

**Keywords** Stress corrosion cracking (SCC) • X-65 • Slow strain rate test (SSRT) • Electrochemical polarization • Sodium bicarbonate solutions

---

M.A. Espinosa-Medina (✉) • G. Carbajal-De la Torre
Facultad de Ingeniería Mecánica, UMSNH, 58000 Morelia, México
e-mail: marespmed@gmail.com; georginacar@gmail.com

C. Ángeles-Chavez
Instituto Mexicano del Petróleo, 07730 México D. F., México
e-mail: cangeles@imp.mx

J.G. González-Rodriguez
Centro de Investigación en Ingeniería y Ciencias Aplicadas, UAEM, 62210 Cuernavaca, México
e-mail: ggonzalez@uaem.mx



R. Pérez Campos et al. (eds.), *Materials Characterization*,
DOI 10.1007/978-3-319-15204-2_10

## 10.1 Introduction

Many kind of materials which could be in contact with several environmental conditions are present in pipeline facilities, where the pressure of process, temperature, fluid transported are additional parameters that interact in conjunction with corrosion attack of materials. Some of those conditions promote the stress corrosion cracking (SCC) damage, that in some cases, acts in association with hydrogen diffusion into the bulk too. Embrittlement by hydrogen increases the structural damage possibility of material under SCC conditions; in this sense, many authors have studied the hydrogen embrittlement susceptibility on different welded stainless steel materials [1–3] and pure and alloyed materials [4–10], where the effects of hydrogen ingress were: loss of plasticity and mechanical resistance, independently of material microstructure.

For the petroleum transport pipeline facilities, there is high interest over the underground pipeline SCC damage possibilities, in this sense, some researches have carried out SCC studies in alkaline solutions [10–16] and alkaline solutions with $H_2S$ and $CO_2$ additions [9]. In near-neutral-pH solutions Parkins and Beavers [10] found evidence of both ingress of hydrogen and metallic dissolution occurring simultaneously. Chen et al. [11] suggested that, pit growing promoted by hydrogen embrittlement as the first step, followed by the anodic dissolution—as an alternative mechanism of the near-neutral pH SCC—occurred on pipeline steel during service. However, the effect of cyclic loading on X-65 and X-80 pipeline steels prior to corrosion exposure (in near-neutral-pH soil environment also) upon the initiation on SCC was observed as microcrack formations on the specimen surface under constant load application [12]. Similarly, Zhang et al. [13] found a comparable behavior of API X-60 in NS4 solution under cyclic and static load conditions, observing similar amount of cracks and a common transgranular stress corrosion crack (TGSCC) characteristics as quasi-cleavage fracture mode [14, 15], mixed with microvoid coalescence. Although pipeline is cathodic protected, as the X-52 pipeline steels in NS4 solution usually do; when protective coating fails, metal dissolution was present promoting crack initiation by hydrogen ingress [16]. However, those results before described were obtained on pipeline body and did not include the welding zone; the goal of this work is to study the susceptibility to SCC and hydrogen permeation effect over weldments of X-65 pipeline steel in diluted $NaHCO_3$ solutions, with and without the presence of $Cl^-$ and $SO_4^{2-}$ ions.

## 10.2 Experimental

Samples of welding zone of X-65 pipeline steel were used for SCC evaluations. Chemical composition is shown in Table 10.1. Smooth tension specimens in accordance with the NACE Standard TM-0198 [17] were machined (with the welded zone in the gauge section) for the environmentally assisted cracking (EAC) evaluations [18], Fig. 10.1a. Tensile specimens were subjected to slow strain rate tests

**Table 10.1** Chemical composition measured from API X-65 pipeline steel

| Element | C[a] | Mn[a] | Si[a] | P[b] | S[b] | Al[b] | Nb[b] | Cu[a] | Cr[a] | Ni[a] | V[b] | Ti[b] |
|---|---|---|---|---|---|---|---|---|---|---|---|---|
| Wt.% | 7 | 146 | 25 | 12 | 2 | 42 | 51 | 10 | 2 | 10 | 70 | 18 |

[a]*x*/100
[b]*x*/1,000

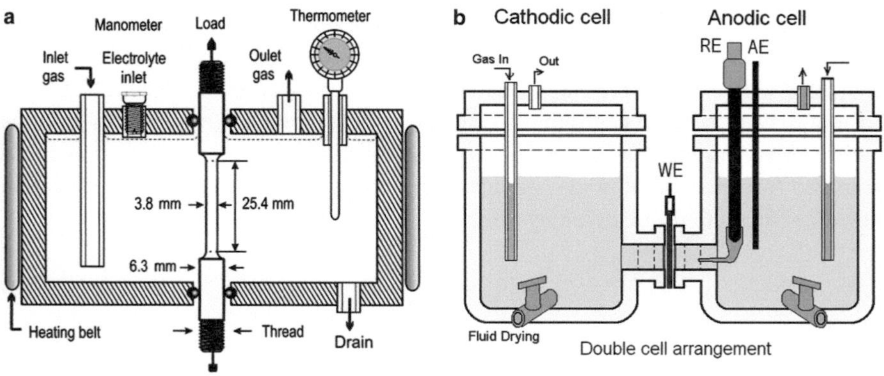


**Fig. 10.1** Schemes of (**a**) SSRT cell setup, and (**b**) double-cell of Devanathan and Stachurski setup

(SSRT) in air, and in 0.1, 0.01, 0.001, and 0.0001 M $NaHCO_3$ solutions at a strain rate of $1.36\times10^{-6}\ s^{-1}$ at 50 °C. All the tests were done at the open circuit potential. The susceptibility to SCC was expressed in terms of the reduction in area percentage (RA%) [17] and the reduction in area ratio (RAR) [17, 18] as follows:

$$\mathrm{RA}(\%) = \frac{\left(D_\mathrm{I}^2 - D_\mathrm{F}^2\right)}{D_\mathrm{I}^2}\times 100 \tag{10.1}$$

$$\mathrm{RAR} = \frac{\mathrm{RA_E}}{\mathrm{RA_A}} \tag{10.2}$$

where $D_\mathrm{I}$ = initial gauge section diameter, $D_\mathrm{F}$ = final gauge section diameter at fracture section, $\mathrm{RA_A}$ = reduction in area in air, $\mathrm{RA_E}$ is the reduction in area in test environment. RAR values near unity generally indicate high resistance to EAC, whereas low values generally indicate low resistance to EAC [17, 18]. The fracture surfaces were then characterized by scanning electron microscopy (SEM) and energy dispersive spectroscopy (EDS). Additional evaluations were realized using solutions with additions of KCl (0.122 g/L) and $MgSO_4$ (0.131 g/L) to the 0.01 M $NaHCO_3$ solution (which presented the highest current density and activity) were done in order to incorporate $Cl^-$ and $SO_4^{2-}$ ions to obtain an approximate chemical concentration of simulated groundwater solution, close to the NS4 solution [16, 19]. Potentiodynamic polarization curves were performed at a sweep rate of 1 mV/s between a polarization window from −1.0 to +1.5 V vs. saturated calomel electrode (SCE), using a fully automated potentiostat controlled with a computer. Electrochemical cell was composed by X-65 sample, SCE and a graphite rod as working, reference and auxiliary electrodes respectively.

Hydrogen permeability measurements were carried out using the double-cell of Devanathan and Stachurski [20], Fig. 10.1b. The 0.1 M NaOH solution was employed at the anodic cell. The specimens were prepared to obtain 0.3 mm thickness plates, polished up to 600 grade emery paper, cleaned with acetone and dried. One face of the thin plate was coated with Pd, which was set on the anodic cell. The anodic cell was filled with oxygen-free NaOH solution and polarized at +250 mV until a passive current density became negligible small. Then the test solution was poured into the hydrogen-generating side to start hydrogen permeation tests. The test solutions were prepared from analytical grade chemicals and distilled water. Hydrogen permeation flux ($J_{ss}$) calculations are given by [20]:

$$J_{SS} = \frac{i_{ss}}{(n)(F)} \tag{10.3}$$

where $J_{ss}$ is the steady state atomic hydrogen permeation in mol H/s/cm$^2$, $i_{ss}$ the steady state atomic hydrogen permeation current density (A/cm$^2$), $F$ the Faraday's constant (96,500 C), and $n$ is the number of transferred electrons for hydrogen.

## 10.3 Results and Discussion

Microstructures of base metal and welding zone of X-65 pipeline steel are showed in Fig. 10.2. The base metal was formed of both equiaxial ferrite grains (*f*) and minority equiaxial perlite grains (*p*) homogenously distributed in the ferrite phase (Fig. 10.2a). The welding zone showed heterogeneous microstructure composed of columnar ferrite grains and fine perlite grains (Fig. 10.2b).

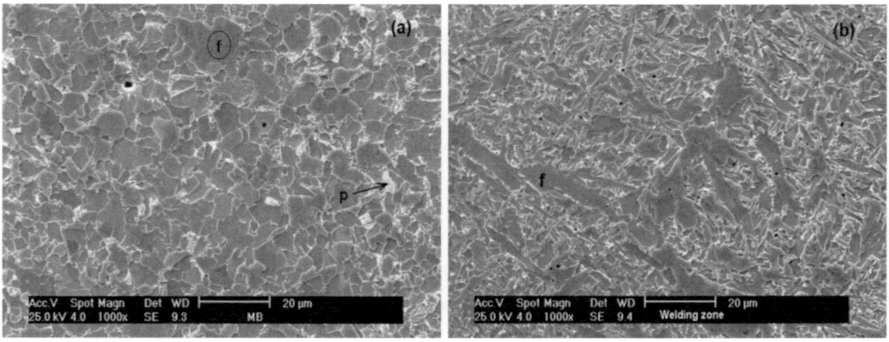


**Fig. 10.2** SEM images of microstructure of X-65 pipeline steel: (**a**) base material, (**b**) longitudinal welding. *f* ferrite, *p* perlite

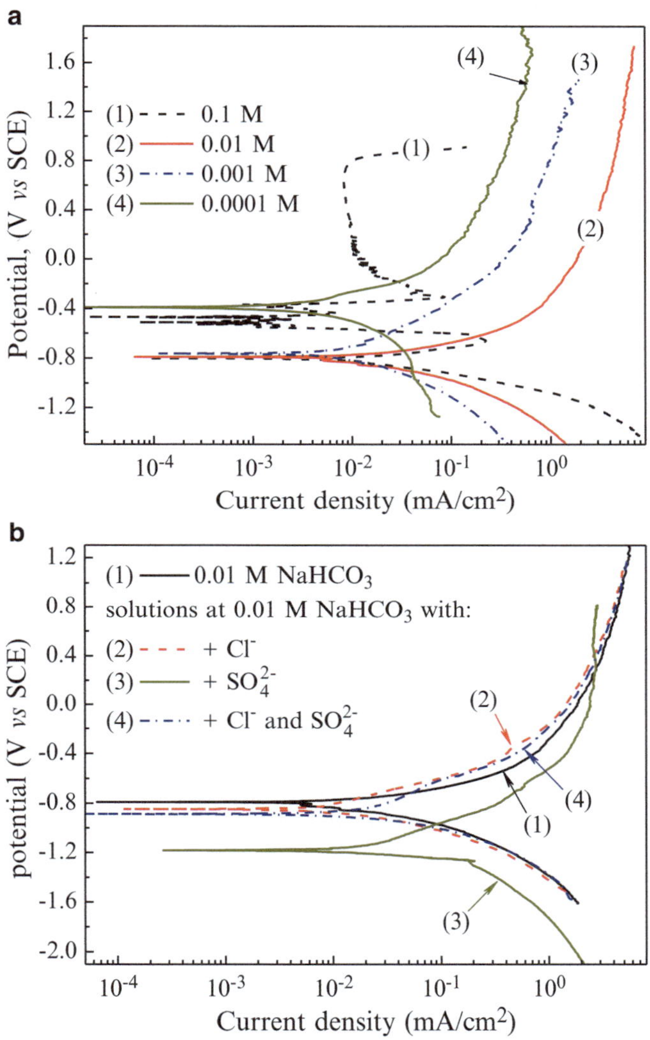


**Fig. 10.3** Potentiodynamic curves of longitudinal weldment of X-65 pipeline steel, in: (**a**) $NaHCO_3$ solutions, and (**b**) 0.01 M $NaHCO_3$ solution with $Cl^-$ and $SO_4^{2-}$ ions additions

Electrochemical polarization results obtained on the welding zone in the solutions with and without chloride and sulfate ions are shown in Fig. 10.3. The effect of $NaHCO_3$ concentration on the potentiodynamic results is shown in Fig. 10.3a. At 0.1 M concentration, a passive region around of 1.0 V was showed, as reported before [15, 21]. Due to the $HCO_3^-$ ion species in that solution and Fe is the main alloying element of pipeline steel, stable ferrous carbonate ($FeCO_3$) film was easily formed according to the following reaction:

$$Fe + HCO_3^- \rightarrow FeCO_3 + H^+ + 2e \quad (10.4)$$

**Table 10.2** Electrochemical polarization parameters measured from curves of Fig. 10.3

| Solution | $E_{corr,}$ $V_{SCE}$ | $i_{corr}$ (mA/cm$^2$) | $\beta a$ (mV) | $\beta c$ (mV) | CR (mm/year) |
|---|---|---|---|---|---|
| 0.1 M $NaHCO_3$ | −0.472 | 0.001 | 44 | 45 | 0.08 |
| 0.01 M $NaHCO_3$ | −0.792 | 0.012 | 113 | 191 | 0.45 |
| 0.001 M $NaHCO_3$ | −0.757 | 0.006 | 263 | 239 | 0.15 |
| 0.0001 M $NaHCO_3$ | −0.393 | 0.002 | 176 | 146 | 0.08 |
| 0.01 M $NaHCO_3$+$Cl^-$ | −0.846 | 0.007 | 192 | 129 | 0.15 |
| 0.01 M $NaHCO_3$+$SO_4^{2-}$ | −1.184 | 0.010 | 139 | 64 | 0.52 |
| 0.01 M $NaHCO_3$+$Cl^-$+$SO_4^{2-}$ | −0.889 | 0.014 | 266 | 138 | 0.20 |

In accord with the reaction, a protective film grows; increasing corrosion resistance at that condition [15]. The passive region was observed from −0.3 to +0.7 V presenting a passive current density, $i_{pass}$ of 0.009 mA/cm$^2$ value approximately, as described before, with the X-52 and X60 steel [21]. At lower $HCO_3^-$ concentration, as the 0.01 M solution, weldment sample showed the more negative $E_{corr}$ (−0.79 V) and higher $i_{corr}$ (0.012 mA/cm$^2$) than obtained at other concentration, however passivation behavior was not observed; it was associated with the pH decreasing. Although the highest anodic current density was showed at 0.01 M, it decreased with the lower solution concentrations [21]. Electrochemical polarization parameters obtained from polarization measurements are showed in Table 10.2. The corrosion rate estimations (CR) at the initial time were calculated using Tafel extrapolation and Faraday's law. Additional polarization measurements were realized using solutions with 0.01 M solution with additions of $Cl^-$ and/or $SO_4^{2-}$ ions (Fig. 10.3b), as described above. The addition of $Cl^-$ or $Cl^-$+$SO_4^{2-}$ ions did not cause any significant change in the polarization response; however, the addition of $SO_4^{2-}$ ions only promoted the $E_{corr}$ displacement to more negative potential (−1.18 V), as reported before [21]. However, the highest corrosion rate was obtained in the 0.01 M solution, when neither $Cl^-$ nor $SO_4^{2-}$ ions were present, but when they were added, the highest corrosion rate was shown in the solution with $SO_4^{2-}$ ions, but it was slightly higher than the one obtained in 0.01 M solution. At the higher $HCO_3^-$ concentration, pitting possibility was small due to passive film formation, but increased at 0.01 M concentration promoting fracture mode by pitting initiation, suggested defective film formation because of low-pH solution. However, addition of $Cl^-$ and $SO_4^{2-}$ ions, caused a decreasing in pH solution, which promoted hydrogen evolution and an increasing on concentration, allowing hydrogen ingress into the metallic substrate.

The hydrogen permeation current density ($J_{ss}$) results are shown in Fig. 10.4. The effect of the solution concentration and additions of $Cl^-$ and $SO_4^{2-}$ on the steady state hydrogen permeation current density, $J_{ss}$ showed its maximum value into 0.01 M $NaHCO_3$ followed by obtained by the immersion of weldment into 0.1 M concentration. An important point is that according to Table 10.2, the highest corrosion rate value for the welding zone was observed at the 0.01 M concentration solution showing higher current density associated with the anodic reaction of $FeCO_3$ formation and promoting an increase in the $H^+$ concentration. Then the maximum

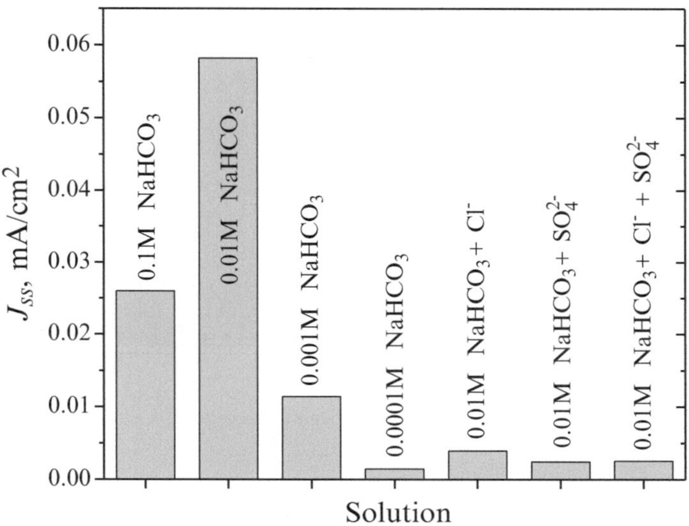


**Fig. 10.4** Effect of $NaHCO_3$ solution concentration and addition of $Cl^-$ and $SO_4^{2-}$ ions on the steady state hydrogen permeation current density, $J_{ss}$, of X-65 weldments

$J_{ss}$ value was found. The effects of $Cl^-$ and $SO_4^{2-}$ ions on the $J_{ss}$ values, obtained for the welding zone in 0.01 M $NaHCO_3$ are shown in Fig. 10.4 also. The addition of those ions in that solution promoted, in all cases, decrease in the $J_{ss}$ values measured at the anodic cell side, mainly with the addition of $SO_4^{2-}$ ions. Although $J_{ss}$ values measured at the anodic side of metallic membrane in the double-cell setup (Fig. 10.1b) were lower, both $SO_4^{2-}$ and $Cl^-$ ions additions promoted the hydrogen permeation within the weldment, but entrapping hydrogen into the sites of microstructure occurred, presenting lower hydrogen permeation flux at the anodic side [22]. Thus, it was suggested hydrogen embrittlement and a concentration of hydrogen adsorption at the crack tip (at bottom of the surface pitting) as effects on the EAC of weldments.

Figure 10.5 shows the effect of the concentration of $NaHCO_3$ solution and additions of $SO_4^{2-}$ and/or $Cl^-$ ions to the 0.01 M $NaHCO_3$ solution in the area reduction percentage (RA%) values (Fig. 10.5a) and RAR values (Fig. 10.5b). It can be seen the lowest RA% was observed in 0.01 M $NaHCO_3$ solution, with 49 % in RA and 0.58 in RAR, while in the other concentrations the RA% was around of the 80 % with RAR values near to the 0.9. However, the addition of $SO_4^{2-}$ or $Cl^-$ ions at the 0.01 M solution caused a decreasing in the RA% around of the 22 %, showing the lowest RAR value (of 0.26) as a clear effect in the SCC susceptibility, indicating low resistance to EAC [17, 18]. Although hydrogen permeation flux quantity was lower at that condition, it was associated with a hydrogen embrittlement effect in the fracture mechanism (Fig. 10.6b), whereas, in 0.01 M $NaHCO_3$ the lower RA% obtained was associated with the hydrogen evolution and its increasing concentration of hydrogen adsorbed at the crack tip (at bottom of the surface pitting) [22],

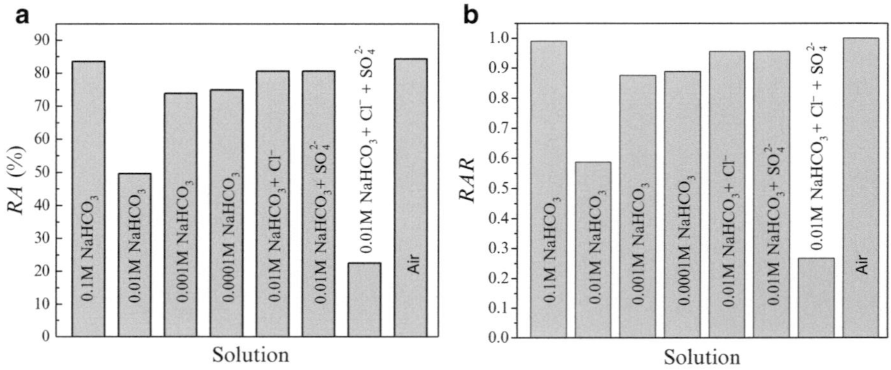


**Fig. 10.5** Effect of $NaHCO_3$ solution concentration and addition of $Cl^-$ and $SO_4^{2-}$ ions on the (**a**) RA(%) and (**b**) RAR, of X-65 weldments

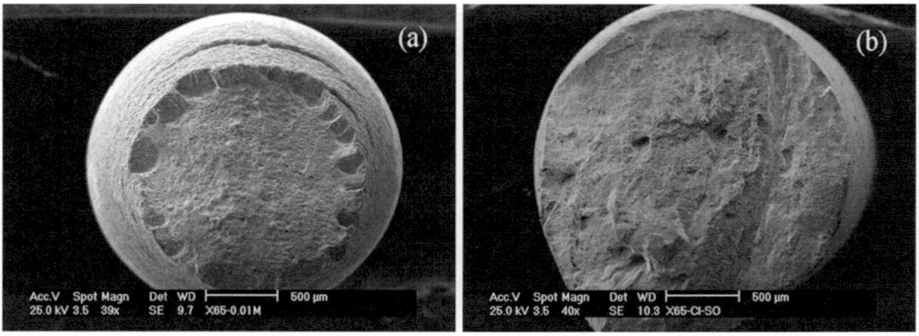


**Fig. 10.6** SEM images of cross section of fractured surface of X-65 weldments in: (**a**) 0.01 M $NaHCO_3$ solution and (**b**) 0.01 M $NaHCO_3$ with $Cl^-$ and $SO_4^{2-}$ ions additions

in accord with $J_{ss}$ measurements (Fig. 10.4) and observed fracture mode (Fig. 10.6a), as reported before [21].

A micrograph of the specimen fracture immersed in 0.01 M $NaHCO_3$ solution is shown in Fig. 10.6a, whereas Fig. 10.6b shows the fracture surface of the specimen strained in 0.01 M $NaHCO_3$ with addition of $Cl^-$ and $SO_4^{2-}$ showing that there is a synergistic effect of these ions, since they alone do not induce significant effect on the EAC of the steel. The fractured surface morphology of those specimens showed brittle fracture characteristic with low RA% and RAR values, while the other specimens showed ductile fracture mode (which are not showed here).

Figure 10.7 shows details of microcracks of fractures showed in Fig. 10.6 at higher magnifications. In 0.01 M $NaHCO_3$ solution, the hydrogen permeation flux ($J_{ss}$) was higher as a cause of higher hydrogen evolution near the anodic interface, followed by an increase in the concentration of hydrogen adsorbed at the surface

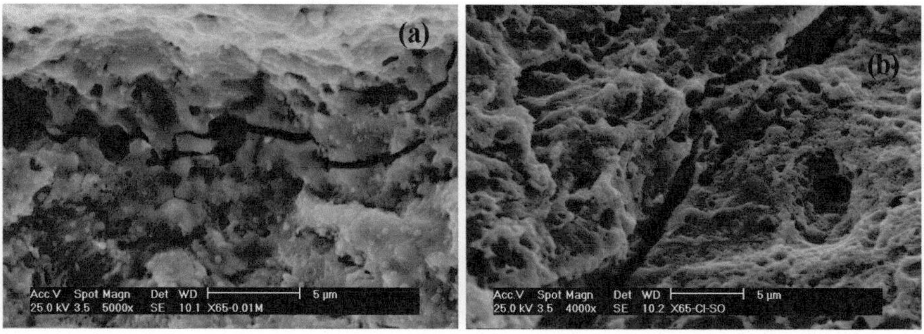


**Fig. 10.7** SEM images of fractured weldments surface showing microcracks in (**a**) 0.01 M $NaHCO_3$ and (**b**) 0.01 M $NaHCO_3$ with $Cl^-$ and $SO_4^{2-}$ solutions

defects, as the crack tips. That promoted the initiation and growing of intergranular crack paths (Fig. 10.7a), as an effect of solution concentration on the EAC of steel [11, 12]. However, in 0.01 M $NaHCO_3$ solution with $Cl^-$ and $SO_4^{2-}$ additions, the quantity of hydrogen permeation flux was modified by the effect of those ions additions, as was showed in Fig. 10.4. But, the corrosion rate of X-56 weldment was increased, causing higher hydrogen generation followed by its permeation into the microstructure [21] promoting an increase in the entrapping hydrogen density as precursor of hydrogen embrittlement mechanism. Therefore, pitting and hydrogen entrapping were the principal effects on the SCC behavior. The fracture mode in that condition was evidenced as transgranular crack paths evolution (Fig. 10.7b) [13].

Because of the weldments showed the highest corrosion rate in the solutions of both 0.01 M $NaHCO_3$ and 0.01 M $NaHCO_3$ with $SO_4^{2-}$ ions, the hydrogen uptake was the highest, which caused an increasing of the SCC susceptibility, similar form than observed in the 0.01 M $NaHCO_3$ + KCl + $Mg_2SO_4$ solution, in accord with their RA% and RAR values. Thus, a fracture mechanism of steel based only in hydrogen absorption and hydrogen embrittlement does not seem to explain the cracking mechanism of X-65 weldments in diluted $NaHCO_3$ solutions at the all concentrations. A fracture mechanism where anodic dissolution plays the most important role is most likely to explain it. The requirements for SCC based on the hydrogen embrittlement mechanism include a susceptible microstructure, a threshold level of hydrogen to induce cracking, and an applied or residual stress [23]. It can be assumed that the failed round tensile bars used in the present study expose every weld microstructure directly to the test solution. In this study, environmental factors, which enhanced hydrogen uptake by the welds, did not also enhance the SCC susceptibility in all cases, but it did enhance the corrosion rate. Thus, the tensile tests showed that these welds are highly susceptible to SCC (Fig. 10.5), because, in air the failure was completely ductile, whereas in the testing solution the fracture mode was very brittle, with very small percentage reduction in area values for two solution conditions, showing several microcrack evidences.

## 10.4 Conclusions

Results of the effects of the $NaHCO_3$ concentration and the additions of $Cl^-$ and $SO_4^{2-}$ ions of the solution at 50 °C, on the corrosion rate, hydrogen permeation and the SSC susceptibility of the X-65 weldments of X-65 pipeline steel, are described as follow:

The CR was highest in 0.01 M $NaHCO_3$ with additions of $SO_4^{2-}$ ions, followed by the immersion in the 0.01 M $NaHCO_3$ solution, while at the 0.1 M $NaHCO_3$ concentration CR was lowest, due to the passive $FeCO_3$ formation.

The SCC susceptibility associated with the %RA and RAR values was higher in the immersion of samples into 0.01 M $NaHCO_3$ and 0.01 M $NaHCO_3$ with $Cl^- + SO_4^{2-}$ ions additions.

The above results presented the feasibility of SCC and hydrogen embrittlement damage of underground pipelines, in metallic sites exposed to solutions with high $HCO_3^-$ and $Cl^- + SO_4^{2-}$ ions concentrations, due to the coating failures and/or cathodic overprotection levels.

**Acknowledgments** The authors thank the CONACYT for the financial support through the extraordinary financial project directed to the Postgraduate Programs of PNPC.

## References

1. Pan C, Su YJ, Chu WY, Li ZB, Liang DT, Qiao LJ (2002) Hydrogen embrittlement of weld metal of austenitic stainless steels. Corros Sci 44:1983–1993
2. Chiang WC, Pu CC, Yu BL, Wu JK (2003) Hydrogen susceptibility of 17-4 PH stainless steel. Mater Lett 57:2485–2488
3. Luu WC, Liu PW, Wu JK (2002) Hydrogen transport and degradation of a commercial duplex stainless steel. Corros Sci 44:1783–1791
4. Danielson MJ (2002) Use of the Devanathan–Stachurski cell to measure hydrogen permeation in aluminum alloys. Corros Sci 44:829–840
5. Tsay LW, Lee WC, Luu WC, Wu JK (2002) Effect of hydrogen environment on the notched tensile properties of T-250 maraging steel annealed by laser treatment. Corros Sci 44:1311–1327
6. Durairajan A, Krishnieyer A, Haran BS, White RE, Popov BN (2000) Characterization of hydrogen permeation through a corrosion-resistant Zinc-Nickel-Phosphorus alloy. Corrosion 56:283–288
7. Kolat VS, Bayri N, Atalay S (2002) Hydrogen effect on the field dependence of Young's modulus of FeCuNbSiB alloy. J Alloys Comp 343:234–237
8. Abd Elhamid MH, Ateya BG, Weil KG, Pickering HW (2001) Effect of Thiosulfate and Sulfite on the permeation rate of hydrogen through Iron. Corrosion 57:428–432
9. Wang SH, Luu WC, Ho KF, Wu JK (2002) Hydrogen permeation in a submerged arc weldment of TMCP steel. Mater Chem Phys 77:447–454
10. Parkins RN, Beavers JA (2003) Some effects of strain rate on the transgranular stress corrosion cracking of ferritic steels in dilute near-neutral-pH solutions. Corrosion 59:258–273

11. Chen W, King F, Vokes E (2002) Characteristics of near-neutral-pH stress corrosion cracks in an X-65 pipeline. Corrosion 58:267–275
12. Wang SH, Chen W, King F, Jack TR, Fessler RR (2002) Precyclic-loading-induced stress corrosion cracking of pipeline steels in a near-neutral-pH soil environment. Corrosion 58:526–534
13. Zhang XY, Lambert SB, Sutherby R, Plumtree A (1999) Transgranular stress corrosion cracking of X-60 pipeline steel in simulated ground water. Corrosion 55:297–305
14. Brongers MPH, Beavers JA, Jaske CE, Delanty BS (2000) Effect of hydrostatic testing on ductile tearing of X-65 line pipe steel with stress corrosion cracks. Corrosion 56:1050–1058
15. Park JJ, Pyun SI, Na KH, Lee SM, Kho YT (2002) Effect of passivity of the oxide film on low-pH stress corrosion cracking of API 5L X-65 pipeline steel in bicarbonate solution. Corrosion 58:329–336
16. Puiggali M, Rousserie S, Touzet M (2002) Fatigue crack initiation on low-carbon steel pipes in a near-neutral-pH environment under potential control conditions. Corrosion 58:961–970
17. NACE Standard TMO198-98 (1988) Slow strain rate test method for screening Corrosion-Resistant Alloys (CRAs) for stress corrosion cracking in sour oilfield service. NACE International, Houston
18. ASTM G 129 (2000) Standard practice for slow strain rate testing to evaluate the susceptibility of metallic materials to environmentally assisted cracking. ASTM, West Conshohocken
19. Torres-Islas A, Serna S, Campillo B, Colin J, Molina A (2013) Hydrogen embrittlement behavior on microalloyed pipeline steel in NS-4 solution. Int J Electrochem Sci 8:7608–7624
20. Devanathan MAV, Stachurski Z (1962) The adsorption and diffusion of electrolytic hydrogen in Palladium. Proc R Soc London A270:90–102
21. Gonzalez-Rodriguez JG, Espinosa-Medina MA, Angeles-Chavez C, Zeferino-Rodriguez T (2007) SCC of X-52 and X-60 weldments in diluted $NaHCO_3$ solutions with chloride and sulfate ions. Mater Corros 58:599–603
22. Griffiths AJ, Thurnbull A (1997) Hydrogen uptake and cracking in 22 % Cr duplex stainless steel under galvanic coupling conditions. Corrosion 53:700–704
23. Kou S (1987) Welding metallurgy. Wiley-Interscience, New York

# Chapter 11
# Electrochemical Characterization of X60 Steel Exposed to Different Soils from South of México

**L.M. Quej, M.J. Míreles, R. Galvan-Martinez, and A. Contreras**

**Abstract** Physicochemical effect on the corrosion process of API X60 steel exposed to five types of soils from south of México at different moisture content was carried out. Two soils were collected in the state of Veracruz (clay of high plasticity and silt) and three soils from the state of Campeche (clay, sand, and clay-silt). Moisture values were determined by addition of 0, 20, 40, and 60 mL of deionized water in a volume of 125 $cm^3$ of each soil. Physicochemical effect of the different soils on the corrosion process of X60 steel was evaluated through electrochemical impedance spectroscopy (EIS) and potentiodynamic polarization curves (PCs). Effect of the damage on the coating when the steel is exposed to corrosive soils was studied. The three more corrosive soils (clay, sand, and clay-silt) were used to evaluate the corrosion behavior with a viscoelastic coating with a simulated damage. The higher corrosion rate obtained from polarization curves for uncoated X60 steel was found in the clay (0.379 mm/year) with 39.7 wt.% of moisture content. Steel coated with a simulated damage indicates that the clay-silt (52.3 wt.%) was the most aggressive soil for X60 steel (0.0029 mm/year).

**Keywords** X60 steel • Soils • Corrosion • Electrochemical impedance spectroscopy (EIS) • Potentiodynamic polarization curves

L.M. Quej • M.J. Míreles • A. Contreras (✉)
Instituto Mexicano del Petróleo, Eje Central Lázaro Cárdenas Norte No. 152, Col. San Bartolo Atepehuacan, Del. Gustavo A. Madero, México 07730, México
e-mail: lquej@imp.mx; mmireles@imp.mx; acontrer@imp.mx

R. Galvan-Martinez
Unidad Anticorrosión, Instituto de Ingeniería, Universidad Veracruzana, Av. S.S. Juan Pablo II s/n, Fracc. Costa Verde, 94294 Veracruz, México
e-mail: rigalvan@uv.mx

 
R. Pérez Campos et al. (eds.), *Materials Characterization*,
DOI 10.1007/978-3-319-15204-2_11

## 11.1 Introduction

External corrosion and stress corrosion cracking (SCC) phenomena are the main deterioration mechanism of the buried pipelines. The origin of these damages is due to coating failure and inefficient cathodic protection, which can reduce the structural integrity of buried pipelines. To preserve the external integrity, pipeline systems are designed and built with two protection systems: one physical (coating) and other electrochemical (cathodic protection) that can be performed by impressed current or sacrificial anodes.

The various phenomena that damage the external surface of pipelines are developed in a complex process, mainly due to the heterogeneity of the soils [1]. The corrosivity of a specific soil generally is determined by soil type, moisture content, soil resistivity, soluble ion content, pH, oxidation–reduction potential, and the role of microorganism in the soil. Soil properties depend on soil particle size distribution, organic content, mineralogical composition, structure and moisture content, that all of them can change inclusive at short distances. There are some mathematical models that use physical parameters measured in field, such as resistivity, pH, redox potential, and level of cathodic protection, to determine the corrosion susceptibility of pipelines [2]. The model is based upon the concept that damage evolution is an irreversible process. These models provide general guidance for materials selection and for the location and design of underground pipelines.

Deterioration by external corrosion in buried pipelines occurs due to formation of corrosion cells between the metallic structure and the soil [3, 4]. The severity of this deterioration is determined by the level of environmental corrosivity. The investigation and studies on SCC originated by the soils in the current days have become of interest in the oil industry and for the researchers in the area of the preservation of materials. SCC can develop in pipelines under normal operating condition when a coating disbonded and groundwater comes into contact with the outside surface of the pipe.

The concern with the environment is very important and a better understanding of the soil as a corrosive agent becomes necessary to optimize the cathodic protection systems in the pipelines. Steel pipelines corrode in soil by complex electrochemical processes because of a different soil electrolytes nature. The corrosion process of X60 pipeline steel using various aqueous solutions representatives of a one type of soil has been developed [5]. As well as some SCC studies using the X60 steel have been developed [6–10]. On the other hand, some studies about mechanical and environmental effects on SCC of low carbon pipeline steel in a soil were carried out [11]. The electrochemical impedance spectroscopy (EIS) results showed that the highest corrosion of the steel sample was obtained when the highest cathodic over potential was applied. Corrosion phenomena are accentuated by the influence of soil parameters such as resistivity, pH, temperature, moisture content, and chemical composition of electrolytes contained in the soil.

It was found that the presence of hydrogen in the steel changes the fracture mode, passing from the microvoid coalescence to quasi-cleavage, accelerating the fracture process [11]. The non-metallic inclusions in the steel act as trap sites of hydrogen. When a critical concentration of hydrogen is reached the crack propagates, following a transgranular path due to the embrittlement of the steel.

Studies associated with the effect of temperature on corrosion process of carbon steel in contact with saline soils and solutions representative of soils suggest that the effect of depth and type of soil are relevant factors to activate the oxidation of steel [12–14]. Some studies use different representative aqueous solutions of soils that are in contact with metal surfaces in a particular geographical area [5, 15], which are used for the characterization of the external corrosion process in different carbon steel. In this study we use directly the collected samples of soil as a corrosive environment. The synthetic solutions of soils do not correspond to the environmental systems of all the types of soils in each geographic region and environmental conditions. Due to the soils diversity that exist in México, this work is proposed to use the EIS and PC to study the corrosion behavior of the API X60 pipeline with and without coating in contact with five different soils types with different moisture content. Two soils were collected in the state of Veracruz and three soils from the state of Campeche in México.

## 11.2 Experimental

### *11.2.1 Steel Used*

The steel sample used as working electrode was made of API 5L X60 pipeline steel with 42 in. (106.6 cm) in diameter and 0.5 in. (1.27 cm) in thickness. Table 11.1 shows the chemical composition of this steel.

Some of the most important mechanical properties for the API X60 pipeline steel are shown in Table 11.2.

The working electrodes for the electrochemical techniques were machined to obtain steel plates with $2\times2\times0.5$ cm. The working electrode was mechanically polished up to 600 grit SiC paper, cleaned, and degreased with acetone prior to perform the tests.

**Table 11.1** Chemical composition of API X60 pipeline steel (wt.%)

| C | Mn | Si | P | S | Cu | Cr | Ni | Nb | V | Ti | Al | Fe |
|---|---|---|---|---|---|---|---|---|---|---|---|---|
| 0.02 | 1.57 | 0.14 | 0.013 | 0.002 | 0.3 | 0.26 | 0.17 | 0.09 | 0.004 | 0.014 | 0.046 | Bal. |

**Table 11.2** Typical mechanical properties of the API X60 pipeline steel

| Steel | YS (MPa) | UTS (MPa) | *E* (GPa) | EL (%) | HV |
|---|---|---|---|---|---|
| X60 | 467 | 567 | 201 | 40 | 201 |

### 11.2.2 Soils Characterization

Physicochemical and electrical properties of the soils collected in the south of México were characterized. Two soils were collected in Veracruz State and three soils in Campeche State in México. These soils were characterized as was found in the field, after they were prepared in lab. First the soils were dried at 60 °C for 3 days, after that the soils were subjected to grinding in order to homogenize the particle size. Subsequently, a constant volume of 125 $cm^3$ sample of each dried and sieved soil was selected. Then in each soil sample different volumes of deionized water were added (0, 20, 40, and 60 mL), in order to dissolve soluble salts and other ionic species. Subsequently, the physicochemical properties were determined again. Thus, moisture content was determined according to ASTM D4959 [16], using a furnace to dry the soils at 110 °C during 1 h. Redox potential was evaluated according to ASTM G200 [17], as well as using an Ag/AgCl electrode redox potential and pH was measured directly in the soil samples.

### 11.2.3 Electrochemical Evaluations

Electrochemical tests were performed in a cell of acrylic as is shown in Fig. 11.1. Steel plates with 1.13 $cm^2$ of total exposure area from API X60 steel were used as working electrode (WE); a graphite rod was used like the auxiliary electrode (AE); a saturated calomel electrode (SCE) as the reference electrode (RE). API X60 steel samples were exposed to the five soils and it was evaluated by EIS and potentiodynamic polarization

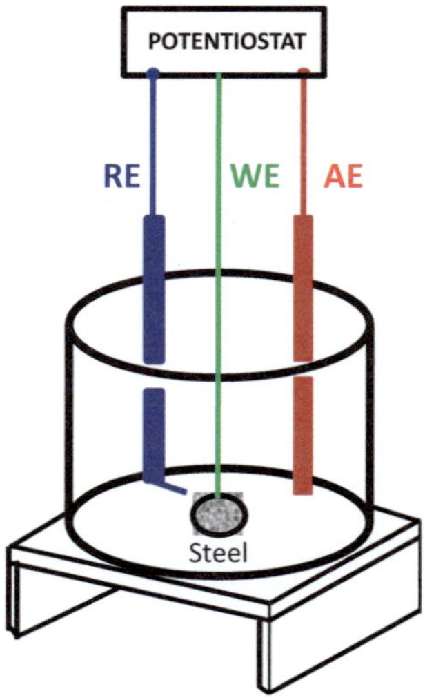


**Fig. 11.1** Experimental setup used to carry out the electrochemical tests

curves (PCs). The impedance spectra were performed in the frequency range of 10 kHz–10 mHz with an amplitude of 10 mV, recording ten points per frequency decade. Potentiodynamic polarization curves were recorded at sweet rate of 0.001 V/s and the polarization potential range used was ±0.3 referred to $E_{corr}$. Electrochemical evaluations with samples uncoated and coated with a Potentiostat–Galvanostat were carried out. Samples were coated with a viscoelastic polymer. The coating has a mechanical damage around 0.3 cm in diameter (0.282 $cm^2$ exposure area).

## 11.3 Results and Discussion

### *11.3.1 Microstructure of the Steel Used*

The API X60 steel microstructure consisted of fine pearlite and ferrite with a grain size around 10–15 μm as is shown in Fig. 11.2. Low carbon steels tend to have a ferrite–pearlite structure containing little pearlite in the grain boundaries mainly. In addition, some globular inclusions containing Mn, S, and Al were observed. It is important to point out that the inclusions are considered as preferred locations for corrosion initiation sites or propagation of pre-existing defects [6–8, 11].

### *11.3.2 Physicochemical Characterization of the Soils*

Figure 11.3 shows images of the five soils collected in south of México. Two soils were collected in the state of Veracruz (clay of high plasticity and silt) and three soils from the state of Campeche (sand, clay, and clay-silt). As can be noted, the soils present colorations and particle size different, which could be associated with changes in the physicochemical properties.

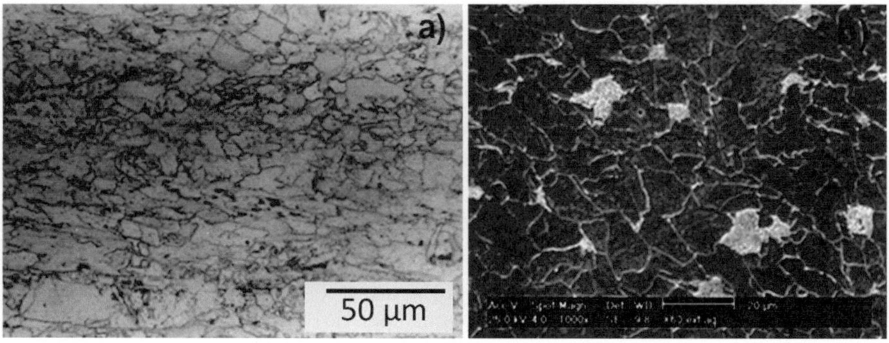


**Fig. 11.2** Typical microstructure of API X60 steel obtained by (**a**) optical microscopy, (**b**) scanning electron microscopy

**Fig. 11.3** Images of the physical view of the soils studied

Table 11.3 shows the physicochemical properties of the soils measured in field directly (0 wt.% of water added) and the properties after 20, 40, and 60 mL of water added. Table 11.3 presents the values of moisture, pH, redox potential, and geographical location of the five soils collected in southern México. The soils evaluated in field show that the moisture content was highest in the clays.

After measurements of the physicochemical parameters of the soils under field conditions (0 wt.% water added), the samples were dried and grinded in order to sieve to have a particle size more homogeneous. From these soils was took 125 $cm^3$ of each soil and was added 20, 40, and 60 mL of deionized water and again physicochemical parameters were measured. The results of these measurements are shown in Table 11.3. Each type of soil absorbs different amount of water. It is observed that most of the soils from the state of Campeche have near neutral pH; soils obtained from Veracruz have pH values with acid trend. The clay from Campeche (Clay-C) provides the maximum value of redox potential (−411 mV) and moisture of 20.35 wt.%. This indicates that the clay should act as real ionic conductor and its behavior could affect metal oxidation.

The moisture, pH, and redox potential presented in Table 11.3 show the complexity of physicochemical phenomena that provide the soils when they are subjected to the presence of water, which indicates that this variation significantly affects the coating and consequently corrosion process of steels.

**Table 11.3** Physicochemical properties of the soils measured in field (0 wt.% water added) and measured in laboratory (adding 20, 40, and 60 mL of water)

| Soil | Location | Water added (mL) | Moisture (wt.%) | pH | Redox potential (mV vs Ag/AgCl) |
|---|---|---|---|---|---|
| Clay of high plasticity (CHP-V) | Near to Zapoapita town (Veracruz) | 0 | 7.02 | 6.5 | −253 |
| | | 20 | 12.22 | 6.65 | −356 |
| | | 40 | 32.11 | 6.69 | −357 |
| | | 60 | 40.75 | 6.75 | −366 |
| Silt (Silt-V) | Near to Coatzacoalcos river (Veracruz) | 0 | 3.43 | 5.5 | −9.35 |
| | | 20 | 21.11 | 6.11 | −288 |
| | | 40 | 29.45 | 5.62 | −375 |
| | | 60 | 36.49 | 5.35 | −399 |
| Sand (Sand-C) | Near to the coast of (Campeche) | 0 | 0.08 | 6.8 | −3.65 |
| | | 20 | 9.82 | 6.79 | −350 |
| | | 40 | 12.80 | 7.05 | −393 |
| | | 60 | 24.92 | 7.27 | −409 |
| Clay (Clay-C) | Near to highway Campeche to Mérida (Campeche) | 0 | 3.63 | 6.4 | −50 |
| | | 20 | 20.35 | 6.22 | −411 |
| | | 40 | 27.85 | 7.06 | −332 |
| | | 60 | 39.71 | 7.67 | −297 |
| Clay-Silt (CS-C) | Near to highway Campeche to Mérida (Campeche) | 0 | 5.48 | 7.2 | −120 |
| | | 20 | 21.04 | 7.21 | −319 |
| | | 40 | 27.42 | 7.35 | −308 |
| | | 60 | 53.12 | 7.75 | −301 |

### *11.3.3 Corrosiveness of the Soils Analyzed by EIS*

Figure 11.4 shows the Nyquist plots obtained for X60 steel exposed to the five types of soils at the maximum water content added. This figure shows that the X60 steel is susceptible to corrosion in these soils. It is important to point out that the impedance spectra were fitting by a simple equivalent electric circuit. According to this it is possible to say that the corrosive effect of soils is as follows: Clay-C>Sand-C>CS-C>Silt-V>CHP-V.

The lowest resistance values correspond to Clay-C and the highest resistance values correspond to CHP-V. This behavior indicates that the corrosion rate (CR) of the X60 pipeline steel is more aggressive in Clay-C than other soils. It is important to point out that the impedance spectrum corresponding X60 pipeline steel in sand of Campeche (sand-C) presents an inductive loop, which it can be associated with the film adsorbed on surface steel. The solution resistance ($R_s$, or conductivity of the soil) in Silt-V and CHP-V is really affected because the values of these $R_s$ are high. This behavior should be attributed to the low concentration of salts in the soils, for that reason they are poor ionic conductors and consequently the CR of the pipeline

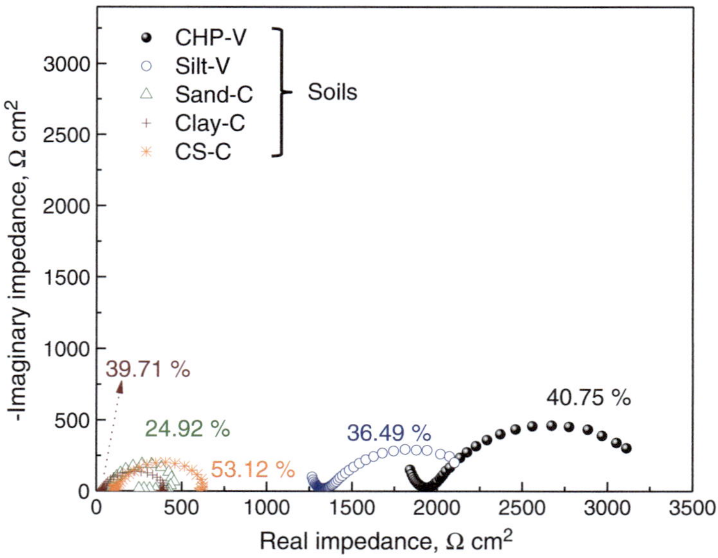


**Fig. 11.4** Nyquist plots obtained for X60 steel exposed to the five soils at the maximum moisture content

steel decreased. In addition, two time constants are possible to observe in these soils, where the time constant at low frequency is attributed to the charge transfer resistance and at high frequency is attributed to the transfer of the oxygen through of the soil.

#### 11.3.3.1 Effect of Moisture Content on Soils

In order to get a better analysis of the soils corrosion on the X60 steel, a depth analysis of the most aggressive soil (Clay-C) was carried out. Figure 11.5 shows the impedance spectra of the X60 steel exposed to Clay-C at three moistures content. It is important to note that the electrochemical phenomenon of Fig. 11.5 reveals the complexity of corrosion process in steel. Because the clays behave as ionic conductors in presence of water, but when the water begins to evaporate (low humidity), this soil will probably not permit the flow of current cathodic protection. In field, this phenomenon could be dangerous because dry clay does not allow adequate cathodic protection, however, when the clay changes from dry to wet clay, it could cause an over potential that would harm the coating and anode beds.

In Fig. 11.5 it is possible to observe that the highest values of real and imaginary impedance for the X60 exposed to Clay-C corresponding to 20.35 wt.% of moisture, this fact indicates that at this moisture, the corrosion rate is low, may be attributed to the soil resistance that limited the corrosion process. However, the moisture value of 39.71 wt.% in the clay activates the corrosion process due to low impedance values.

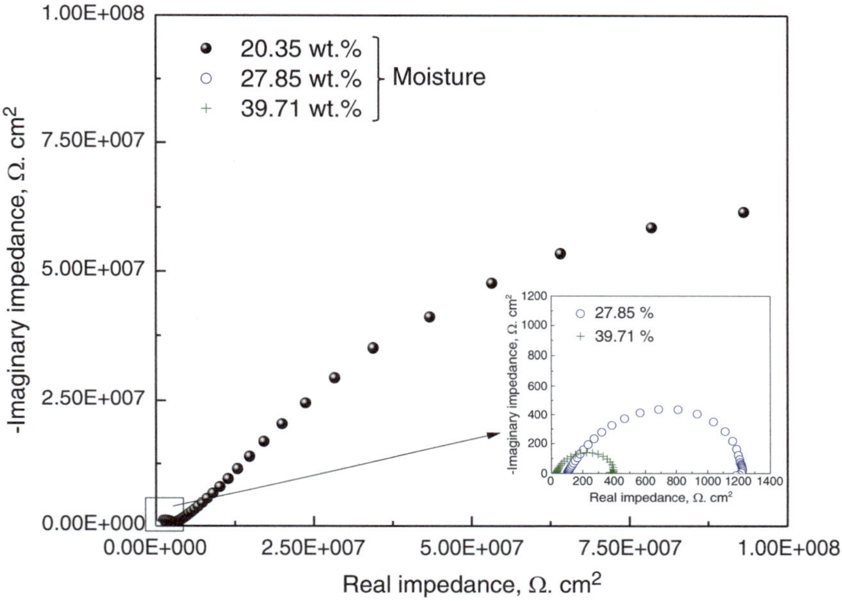


**Fig. 11.5** Nyquist plots obtained for X60 exposed to Clay-C with different moisture content

#### 11.3.3.2 Effect of Coating

The underground pipelines are protected by two forms, one mechanical barrier with a coating and the other electrochemical with galvanic anodes or impressed current. However, there is a high probability of finding damages on the coating with different geometry and morphology (by mechanical damage, bad application or degradation). For that reason it is useful to understand the effect of the damage on the coating when the steel is exposed to a corrosive soil. Thus, were selected the three more corrosive soils (Clay-C, CS-C, Sand-C) to carry out electrochemical evaluations and study of the corrosion process on X60 pipeline steel. Plates from X60 steel were coated and, one damage on the coating was produced (0.3 cm in diameter) prior to electrochemical assessment. The EIS spectra were obtained after X60 steel was exposed 30 days to the soils.

Impedance spectra obtained for X60 steel coated with a commercial coating exposed to the three soils from Campeche are shown in Fig. 11.6. In this figure it is possible to observe that X60 steel is more susceptible to corrosion when it is exposed to CS-C. It is important to mention that the electrochemical phenomena shown in these curves are associated with the response of the induced damage area. Thus, considering the defect on the coating has the same area, the electrochemical changes are attributed to the damage and the loss of the coating adherence due to the presence of these types of soils, so the coating causes different type of corrosion forms likely pitting and crevice corrosion.

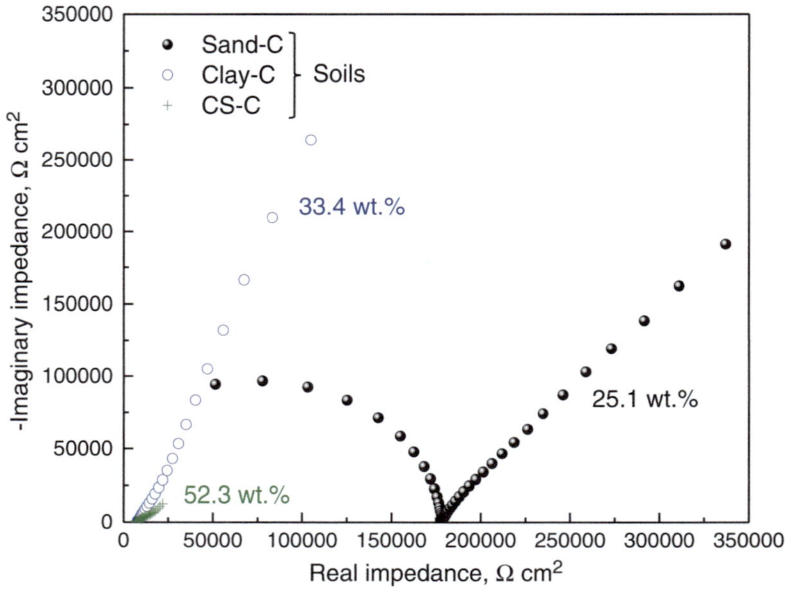


**Fig. 11.6** Nyquist plots obtained for X60 steel coated, with a damage exposed to the three soils from Campeche at the maximum moisture content (after 30 days in soils exposure)

#### 11.3.3.3 Quantitative Analysis of EIS Spectra

In order to carry out a quantitative analysis of the impedance spectra, an adjustment of the experimental data was performed by means of equivalent circuits and the program nonlinear least squares fit of Boukamp [18]. Due to the existence in the majority of the spectra of at least a semicircle, for adjustment of the experimental data an equivalent circuit of a single time constant $R_s$ ($R_{ct}$ $C_{EDL}$) was used, where $R_s$ is the resistance associated with the conductive properties of soil used, $R_{ct}$ is the charge transfer resistance, and $C_{EDL}$ is the capacitance associated with the electrical double layer. The capacitance values shown in Table 11.4 for the properties of the electric double layer were calculated using Eq. (11.1) [19, 20]:

$$C = \frac{\left(Y_o R\right)^{1/n}}{R} \tag{11.1}$$

It is known that depending on the value of capacitance, the active area of the metal is changed [18]. The capacitance assessment ($C$) was carried out in function of the area ($A$), charge ($q$), the dielectric constant ($\varepsilon$), the disturbance voltage ($E$), and the spacing between the parallel flat plates as described in Eqs. (11.2) and (11.3), where $C_i$ is the capacitance value of an $i$ (μF) system; $\varepsilon_o$ is the vacuum dielectric constant ($8.85 \times 10^{-14}$ F cm$^{-1}$); $\varepsilon_i$ is the dielectric constant of the $i$ system; $A$ and $\delta$ correspond to the area and thickness; d$q$ is the change of the charge (Coulomb); d$E$ is the change in potential (Volts).

**Table 11.4** Resistance and capacitance values for X60 steel exposed to soils at different moisture content

| Soil | Moisture content (wt.%) | $R_s$ (Ω) | $R_{ct}$ (Ω cm²) | $C_{EDL}$ (μF/cm²) |
|---|---|---|---|---|
| CHP-V | 12.22 | 2,397 | 7,632 | 1.39 |
| | 32.11 | 1,465 | 2,641 | 114.21 |
| | 40.75 | 1,559 | 3,223 | 98.24 |
| Silt-V | 21.11 | 16,466 | 29,888 | 3.57 |
| | 29.45 | 15,737 | 2,887.4 | 3.51 |
| | 36.49 | 1,335 | 790.02 | 124.77 |
| Sand-C | 9.82 | 2.141 M | 1.8985 G | 8.383 nF |
| | 12.80 | 5,391 | 64.189 M | 8.401 pF |
| | 24.92 | 65.253 | 529.8 | 585.04 |
| Clay-C | 20.35 | 1.5261 M | 5.6134 M | 2.835 pF |
| | 27.85 | 104.29 | 1,139 | 36.119 |
| | 39.71 | 30.16 | 371.78 | 282.58 |
| CS-C | 21.04 | 152,620 | 9.8428 M | 51.89 nF |
| | 27.42 | 634,640 | 89.61 M | 47.47 nF |
| | 53.12 | 100.7 | 572.27 | 247.82 |

*M* mega, *G* giga, *n* nano, *p* pico

$$C_i = \left( \frac{\varepsilon_0 \varepsilon_i A_i}{\delta_i} \right) \tag{11.2}$$

$$\delta \leq C = \left( \frac{dq}{dE} \right) \geq \delta \tag{11.3}$$

If the charge transfer properties retain the same value ε, then the next analysis is carried out:

If $R_{ct}$ is high, then $C_{EDL}$ is low
If $R_{ct}$ is low, then $C_{EDL}$ is high

This indicates that a greater charge transfer resistance ($R_{ct}$) a lower contribution of $C$ is generated, indicating the existence of a lower electrochemical activity in the electrical double layer and a suitable electric field oriented in the steel–soil interface, which control the corrosion process. The greater value of $C$ indicates a greater corrosion due to a lower contribution of the charge transfer resistance in the steel–soil interface.

Table 11.4 shows the values of the electrical parameters obtained by the best fit of the experimental data obtained from the X60 steel exposed to the five soils. According to this table it is observed that X60 steel is more susceptible to corrosion in Clay-C with $R_{ct}$ = 371.78 Ω cm² at a moisture content of 39.71 wt.%. While $C_{EDL}$ values associated with faradaic charge transfer on the Steel Surface indicate that corrosiveness of X60 steel is in the following order: Sand-C, Clay-C, CS-C, Silt-V and finally CHP-V.

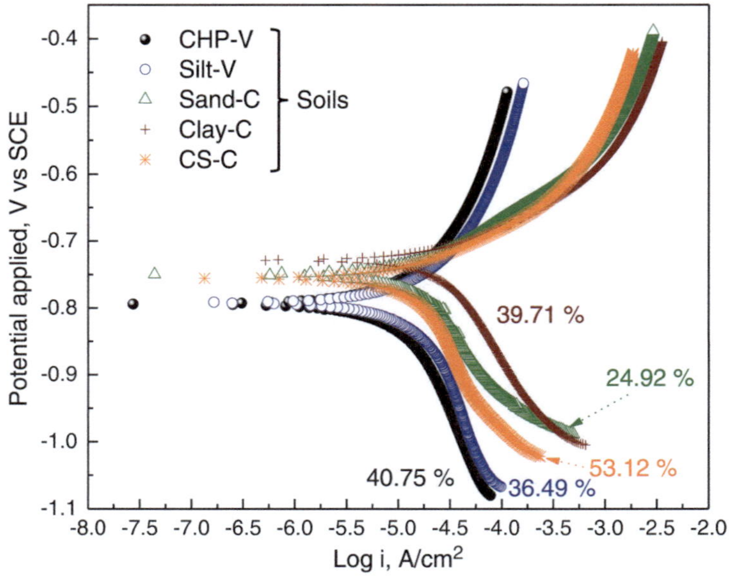


**Fig. 11.7** PCs obtained for X60 steel exposed to the five soils at the maximum moisture content

## *11.3.4 Polarization Curves for Uncoated Steel*

Figure 11.7 shows polarization curves obtained for X60 steel exposed to five soils studied at maximum moisture content (adding 60 mL of deionized water). In this figure, it is observed that in all PCs of the X60 steel samples, the anodic branches have high slopes. This behavior indicates that the oxidation reaction (charge transfer polarization) is influenced by a mass transfer process; it is attributed to the wet soil adsorbed on surface of the steel samples. On the other hand, the cathodic branches in all PCs have high slopes too. This fact indicates that a diffusion process also influenced in the cathodic reaction. According to the behavior observed in the PCs it is possible to mention that the corrosive effects of each soil are in the following order: Clay-C>CS-C>Sand-C>Silt-V>CHP-V.

The electrochemical responses in the qualitative analysis show that there are at least three types of soils those are corrosives to steel, these are the soils from Campeche. Furthermore, it is well known that the external surfaces of the buried steel pipeline often are exposed to different environmental conditions that may change the corrosiveness of the soils.

### 11.3.4.1 Quantitative Analysis of PCs

Table 11.5 shows the corrosion rates values obtained from polarization curves for X60 steel exposed to saturated soils (Fig. 11.7). From these results it is observed that X60 steel is more susceptible to corrosion in Clay-C with a corrosion rate of 0.3796 mm/year; followed by CS-C (0.3475 mm/year), Sand-C (0.3013 mm/year), Silt-V (0.2700 mm/year), and finally the CHP-V (0.1813 mm/year).

**Table 11.5** Corrosion rates values obtained from polarization curves of the X60 steel exposed to soils at the maximum moisture content

| Soil | Moisture content (wt.%) | CR (mm/year) |
|---|---|---|
| CHP-V | 40.75 | 0.1813 |
| Silt-V | 36.49 | 0.2700 |
| Sand-C | 24.92 | 0.3013 |
| Clay-C | 39.71 | 0.3796 |
| CS-C | 53.12 | 0.3475 |

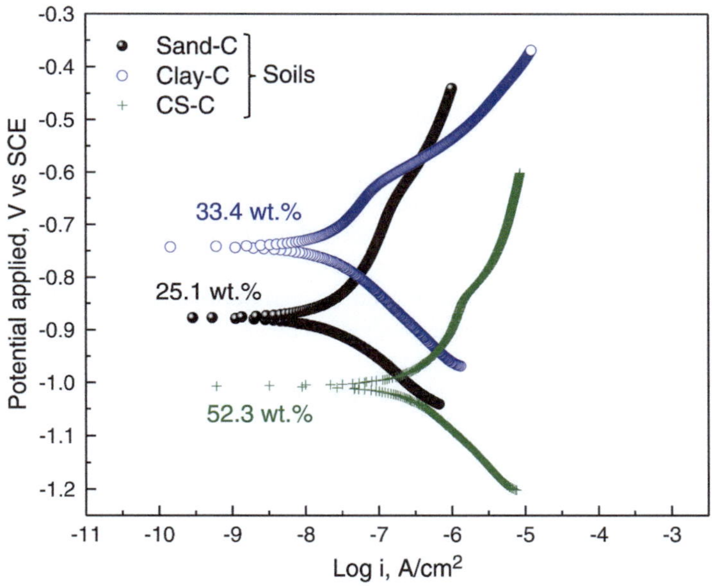


**Fig. 11.8** Potentiodynamic polarization curves obtained for X60 steel coated, with a damage exposed to the soils from Campeche at the maximum moisture content (after 30 days in soils exposure)

The qualitative and quantitative analyses show the complexity of the corrosion process of steels in soils, because each type of soil has different physicochemical properties, such as color, pH, redox potential, corrosion potential, particle size, ions and moisture retention, which must be considered in electrochemical assessment.

### *11.3.5 Polarization Curves for Coated Steel*

Figure 11.8 shows potentiodynamic polarization curves obtained for X60 steel coated with a viscoelastic polymer having a simulated damage and exposed to three soils from Campeche at maximum moisture content (adding 60 mL of water). The clay-silt has a greater retention of water (52.3 wt.%) due to the physicochemical

**Table 11.6** Physicochemical parameters and corrosion rate for X60 steel coated, with damage exposed to the soils from Campeche at the maximum moisture content (after 30 days)

| Soil | Moisture content (wt.%) | Redox potential (mV vs Ag/AgCl) | pH | $E_{corr}$ (mV vs ECS) | CR (mm/year) |
|---|---|---|---|---|---|
| Sand-C | 25.1 | −305 | 6.53 | −876.55 | 0.000198 |
| Clay-C | 33.4 | −280 | 7.47 | −742.14 | 0.000246 |
| CS-C | 52.3 | −285 | 7.21 | −1,006 | 0.002951 |

properties [19, 20]. In addition, this type of soil increases the current values in the branch of the anode and cathode, respectively. This behavior suggested that the CS-C is more corrosive than Sand-C and Clay-C for the X60 steel. It is important to mention that the anodic branches for the X60 steel in the three soils have slopes that belong at not pure charge transfer process. It is to say, a passivation process is influenced in the oxidation reaction of the X60 steel corrosion. It is important to point out that the anodic branches also have a slight decrease in the value of the oxidation current of about 100–120 mV from $E_{corr}$, then after these potential values is observed a significant increase in the current response. The slight decrease in current indicates that the formation of corrosion products and the wet soil absorbed on surface of the steel could have produced a passivation behavior that it can have limited the charge transfer between the steel-coating-soil. In addition, the cathodic branches show the contribution of the diffusion process overall in the CS-C soil.

#### 11.3.5.1 Quantitative Analysis of PCs for Coated Steel

Table 11.6 lists physicochemical parameters and corrosion rate obtained from polarization curves for X60 steel (Fig. 11.8) exposed to the soils from Campeche after 30 days. It is observed that the X60 showed the biggest corrosion rate when is exposed to CS-C, which corroborates with the impedance diagrams of Fig. 11.6. It is observed that Clays in general are more corrosive soil and probably modify the properties of adhesion of the coating to the metal. It is observed that the same coating on the steel acts different, it could be that the different soil modifies its adhesion properties changing the exposure area and the corrosion process, demonstrating the complexity of the corrosion process of steel in contact with different soils even with the presence of coatings.

## 11.4 Conclusions

Physicochemical effect on the corrosion process of API X60 steel exposed to five types of soils from south of México at different moisture content was carried out. The soils evaluated have different water retention and each soil has a moisture content in which the corrosion process is more active. The greater corrosion rate, do not belong necessarily in saturated soils. The higher corrosion rate obtained from polarization curves for uncoated X60 steel was found in the Clay-C (0.379 mm/year) with 39.7 wt.% of moisture content. Steel coated with a simulated damage indicates that the Clay-Silt (52.3 wt.%) was the most aggressive soil for X60 steel (0.00295 mm/year). These results are agreed with the results obtained by EIS. In addition, it is important to point out that a not pure charge transfer process is developed in the corrosion of the X60 steel; it is to say, a diffusion process influenced the activation corrosion process. The results suggest the use of EIS and PC in the cathodic protection study is a really important tool for the analysis of the corrosion process in underground pipelines at different stages of season of the year (dry and rainy), due to the cathodic potential variations in different types of soils and different moisture contents.

## References

1. Wilmott MJ, Jack TR (2000) Corrosion by soils. In: Winston Revie R (ed) Uhlig's corrosion handbook, 2nd edn. Wiley & Sons, Inc., ISBN 0-471-15777-5
2. Alamilla JL, Espinosa-Medina MA, Sosa E (2009) Modeling steel corrosion damage in soil environment. Corros Sci 51:2628–2638
3. Beavers JA, Thomson NG (2006) External corrosion of oil and natural gas pipelines. ASM Handbook, vol 13C, Corrosion: environments and industries (#05145). ASM International, Materials Park, 44073-0002, 1021–1022
4. Uhlig HH (1971) Corrosion and corrosion control. Introduction to corrosion science and engineering, 2nd edn. Wiley, New York
5. Benmoussat A, Hadjel M (2005) Corrosion behavior of low carbon line pipe steel in soil environment. J Corros Sci Eng 7:1–19
6. Velazquez Z, Guzman E, Espinosa MA, Contreras A (2010) Stress corrosion cracking behavior of X60 pipe steel in soil environment, Mater Res Symp Proc 1242:69–78 [Pérez R, Contreras A, Esparza R (eds)]
7. Contreras A, Hernández SL, Galvan R (2011) Effect of pH and temperature on stress corrosion cracking of API X60 pipeline steel. Mater Res Symp Proc 1275:43–52 [Pérez R, Contreras A, Esparza R (eds)]
8. Pan BW, Peng X, Chu WY, Su YJ, Qiao LJ (2006) Stress corrosion cracking of API X-60 pipeline in a soil containing water. Mater Sci Eng A 434:76–81
9. Trejo-Atecas B, Orozco-Cruz R, Contreras A, Galvan-Martinez R (2014) Electrochemical characterization of the X60 steel corrosion at different overvoltages: a cathodic protection study. Mater Sci Forum 793:59–66
10. Quej-Aké LM, Galvan-Martinez R, Contreras A (2013) Electrochemical and tension tests behavior of API 5L X60 pipeline steel in a simulated soil solution. Mater Sci Forum 755:153–161
11. Contreras A, Hernández SL, Orozco-Cruz R, Galvan-Martinez R (2012) Mechanical and environmental effects on stress corrosion cracking of low carbon pipeline steel in a soil solution. Mater Design 35:281–289
12. Nie XH, Li XG, Du CW, Cheng YF (2009) Temperature dependence of the electrochemical corrosion characteristics of carbon steel in a salty soil. J Appl Electrochem 39:277–282
13. Gervasio D, Song I, Payer JH (1998) Determination of the oxygen reduction products on ASTM A516 steel during cathodic protection. J Appl Electrochem 28:979–992
14. Murray JN, Moran PJ (1989) An EIS study of the corrosion behavior of polyethylene coating holidays in natural soil conditions. Corrosion 45:885–895

15. Fitzgerald JH (1993) Evaluating soil corrosivity-then and now. Mater Perform 49:17–19
16. ASTM D 4959 (2007) Standard test method for determination of water (moisture) content of soil by directs heating. ASTM International, West Conshohocken, PA, 2007
17. ASTM G 200-09 (2014) Standard test method for measurement of oxidation-reduction potential (ORP) of soil. ASTM International, West Conshohocken, PA, 2014
18. Boukamp BA (1989) Equivalent Circuit (EQUIVCRT.PAS), User Manual, University of twente, Netherlands
19. Quej-Ake L, Marín-Cruz J, Galván-Martínez R, Contreras A (2014) Corrosion behavior of low carbon steel exposed to different soils. Mater Sci Forum 793:169–179
20. Pech-Canul MA, Chi-Canul LP (1999) Investigation of the inhibitive effect of N-phosphonomethyl-glycine on the corrosion of carbon steel in neutral solutions by electrochemical techniques. Corrosion 55:948–956

# Part IV
# Characterization of Stainless Steels

This part includes two chapters related to the characterization of stainless steels more used in the industry, which include AISI 430 stainless and 304 Austenitic Stainless Steel.

One of the chapters presents results about characterization of the mechanical behavior of AISI 430 steel using the Mapping of the Polycrystalline Flow (MPF), during axial tension test of Spatially Extended Polycrystalline Systems (SEPCS) which allows determining relevant aspects of the phenomenology and mechanics of polycrystalline metals flow and alloys during irreversible deformation.

The second chapter studied the effect of the perpendicular electromagnetic field in the 304 austenitic stainless steel welding process. Plates of AISI 304 with a single small "V" groove configuration in upper side were joined using the gas metal arc welding process (GMAW). A perpendicular electromagnetic field was applied in order to obtain high penetration and to improve the degree of sensitization in the heat affected zone (HAZ). The effect of the magnetic field was evaluated in terms of susceptibility to localized corrosion in the heat affected zone.

The main techniques used in the characterization of these stainless steels are:

- Optical microscopy (OM).
- Scanning electron microscopy (SEM).
- Mapping of the Polycrystalline Flow (MPF).
- Mechanical tests.
- Electrochemical test (double loop-electrochemical potentiokinetic reactivation test).

# Chapter 12
# Effect of the Perpendicular Electromagnetic Field in the 304 Austenitic Stainless Steel Welding in a Single Pass

**Rafael García, Rafael Cortes, Diana L. García, and Víctor H. López**

**Abstract** Plates, 12.7 mm thick, of AISI 304 austenitic stainless steel were welded using a non-conventional joint design in a single pass with and without the application of magnetic fields with different orientations. A perpendicular electromagnetic field was applied in order to obtain high penetration by turning helicoidally the path of the plasma within the groove of the joint during welding. Observation of the transverse profiles of the welds revealed fully penetrated welds; however, lack of sidewalls fusion was seen in every profile. Welding with magnetic field decreased the extent of this defect. The effect of the magnetic field was evaluated in terms of susceptibility to localized corrosion in the heat affected zone (HAZ) by the double loop-electrochemical potentiokinetic reactivation (DL-EPR) test. These results indicated that the as-received plates were already sensitized and plain welding worsened this condition in the HAZ. DL-EPR curves revealed that welding with perpendicular magnetic field reduced the degree of sensitization in the HAZ. Examination of the essayed surfaces in the scanning electron microscope (SEM) showed that generalized corrosion took place instead of localized corrosion.

**Keywords** 304 austenitic stainless steel • Electromagnetic field • Single pass • Joint • Welding

R. García (✉) • R. Cortes • V.H. López
Instituto de Investigación en Metalurgia y Materiales, Universidad Michoacana de San Nicolás de Hidalgo, Morelia, Mich., México
e-mail: rgarcia@umich.mx; crafiux@gmail.com; composito@yahoo.com

D.L. García
Facultad de Ingeniería Química, UMSNH, Morelia, Mich., México
e-mail: united_moon_sun@hotmail.com

 
R. Pérez Campos et al. (eds.), *Materials Characterization*,
DOI 10.1007/978-3-319-15204-2_12

## 12.1 Introduction

The austenitic stainless steels are one of the broader groups of stainless steels with a vast field of applications due to their mechanical properties and corrosion resistance. One of the principal manufacturing processes in the industry is fusion welding where it is very common that this kind of steels experience the sensitization phenomenon in the heat affected zone (HAZ) and this thermal affection makes this region susceptible to localized corrosion owing to chromium depletion around precipitated carbides or Cr-rich compounds. The precipitation of chromium carbides along the grain boundaries takes place between 450 and 850 °C, this temperature range is compulsorily reached by the welding thermal cycle in the HAZ [1–4]. The carbides are often associated with grain boundaries with high values of the relation between the areas of the unit lattice forming coincident sites lattice (CSL) or large disorientation angles ($\Delta\Theta$) for CSL [5]. Thus, in the HAZ of austenitic stainless steels it is very common to occur sensitization, which adversely affects their intergranular corrosion resistance in seawater and in a number of acids.

Nowadays there are available several methods for minimizing the problems that arise when austenitic stainless steels are welded. Thermomechanical treatment before welding increases the fraction of grain boundaries with low disorientation angles; these grain boundaries are of low energy and less likely for chromium carbide precipitation [6]. Superficial partial fusion using laser welding is used to generate a solidification microstructure free of sensitization [7]. In spot welding, the application of large magnetic fields has been employed but no report is given about the effect of this practice with respect to sensitization [8]. Other researchers looked at the effect that has the application of magnetic fields on carbon diffusion. It was found that the perpendicular magnetic field applied is the best in the carbon diffusion, when the steel is in austenitic state [9].

The use of electromagnetic interaction of low intensity (EMILI) is newly implemented in the austenitic stainless steels. When an external magnetic field of low intensity is applied parallel to the electric arc during welding and this magnetic field interacts with the magnetic field produced by the welding current, the EMILI produced leads to an improvement of the intergranular corrosion in the HAZ of the austenitic stainless steels [10–12]. The interaction between two magnetic fields enhances Cr short-range diffusion by vibration of the crystalline lattice and this effect heals Cr decay zones around chromium carbides [11, 12].

This study was focused on the application of magnetic fields of constant intensity, during welding, with three different orientations for determining which of them yields full penetration and an improvement in the corrosion resistance in the HAZ.

## 12.2 Experimental

Plates of AISI 304 ($150\times75\times12.7$ mm$^3$) with a single small "V" groove configuration in upper side, as shown in Fig. 12.1, were joined using the gas metal arc welding process (GMAW) with an ER308L Si filler wire, 1.6 mm in diameter, fed at 170 mm/s.

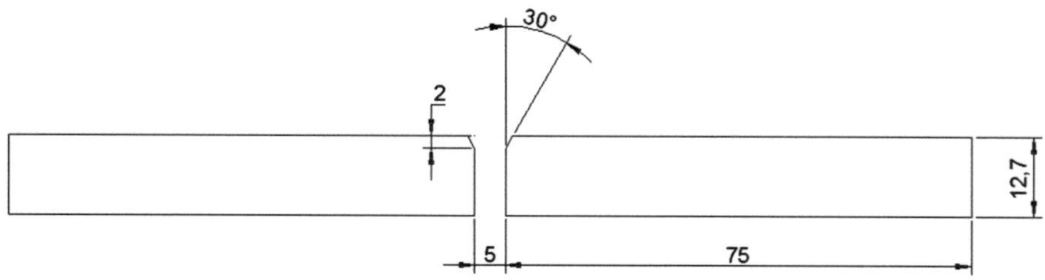


**Fig. 12.1** Scheme and dimensions (mm) of the joint design used in the welding trials

**Table 12.1** Chemical composition of the base metal and electrode (wt.%)

| Material | C | Mn | Si | P | S | Ni | Cr | Mo |
|---|---|---|---|---|---|---|---|---|
| AISI 304 | 0.046 | 1.54 | 0.34 | 0.03 | 0.01 | 8.32 | 18.32 | – |
| ER308L-Si | 0.03 | 1.5 | 0.75 | 0.03 | 0.03 | 12.5 | 23 | 0.75 |

**Table 12.2** Welding parameters

| Welding trial | Voltage (V) | Current (A) | Wire feed (mm/s) | Welding rate (mm/s) | Magnetic field setup |
|---|---|---|---|---|---|
| 1 | 25.3 | 423 | 170 | 2.25 | (a) |
| 2 | 25.5 | 423 | 170 | 2.25 | Without MF |
| 11 | 25.3 | 414 | 170 | 2.25 | (b) |
| 12 | 25.4 | 425 | 170 | 2.25 | (c) |

A shielding gas mixture: 98 % Ar+2 % $O_2$ flowing at 25 L/min was used. This joint design was used in order to test if the perpendicular magnetic field applied, as described later, could produce full penetration in a single welding pass in plates with 12.7 mm in thickness. The chemical composition of the base metal (BM) and filler wire are listed in Table 12.1. The welding torch was displaced at 2.25 mm/s with a stick out of 12 mm. A constant voltage power supply was used with direct current-electrode positive. Welding parameters are shown in Table 12.2. It is worth mentioning that a number of welding trials were performed but for the sake of brevity only four cases are presented here. The numbers listed in the first column and the letters in the last column in Table 12.2 correspond to the welding trial and the experimental setup used for welding (Fig. 12.2), respectively. The external perpendicular magnetic field intensity applied was of 3 mT with three different orientations as shown in Fig. 12.2. The magnetic field value was chosen with reference to previous studies [11, 12]. Optical microscopy (OM) was conducted on samples in the as-received condition and on samples taken from transverse sections of the welded joints after standard metallographic preparation followed by etching in a solution of 0.8%$HNO_3$+2.1%HCl+97.1%$C_2H_6O$ to reveal the welding profiles.

The electrochemical characterization of the HAZ was carried out at 3 mm from the fusion line using the double loop-electrochemical potentiokinetic reactivation test (DL-EPR) to determine the degree of sensitization (DOS), which is determined by the ratio $I_r/I_a$ where $I_r$ is the peak reactivation current and $I_a$ is the peak activation current.

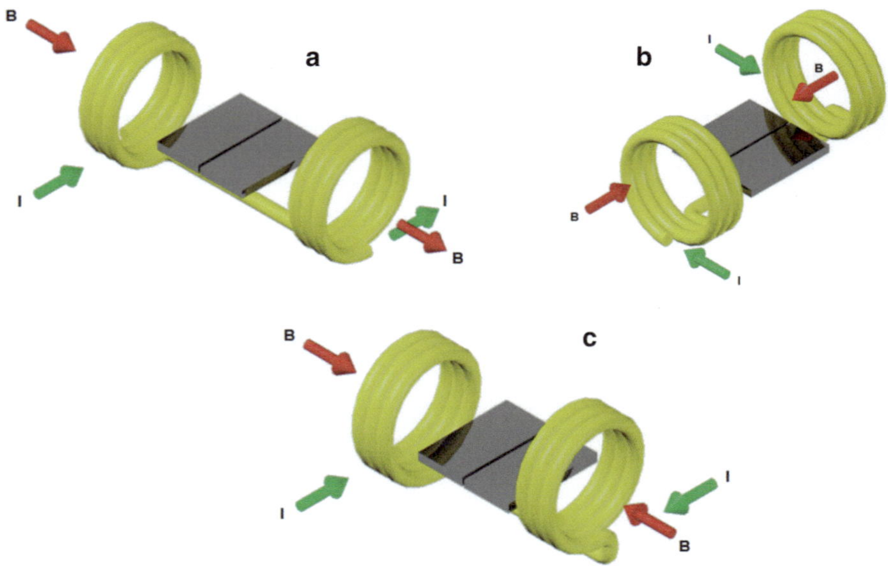


**Fig. 12.2** Schematic of the application of the external magnetic field. (**a**) Perpendicular in one direction (welding trial 1). (**b**) Perpendicular opposite axial (welding trial 11) and (**c**) Perpendicular opposite (welding trial 12). Notice that the direction in which the current, *I*, is fed into the coil leads to different orientations in the magnetic field, *B*, and therefore the electromagnetic interaction with the austenitic steel plates is also different. In setups (**b**) and (**c**), the electromagnetic fields generated with the coils are the same but the groove of the joint has different orientations with respect to the coils

Tests were performed on samples of $10\times8$ mm$^2$ cut from the HAZ of the welds. The samples were embedded in epoxy resin to be used as working electrodes with a copper wire attached to the rear part for electric contact. A conventional three-electrode electrochemical cell was used with the BM or weld samples as working electrode, a saturated calomel electrode (SCE) as reference electrode, and graphite bar as auxiliary electrode. Tests were made three times for reproducibility using samples subjected to identical surface preparation; grinding with SiC paper up to 1,500 grit, rinsing with distilled water and degreasing with acetone. Fresh electrolyte was used for every electrochemical test. The DL-EPR tests were conducted in 0.5 M $H_2SO_4$+0.01 M KSCN solution at room temperature as recommended in ASTM G-108 and reference [13]. The samples were polarized anodically using a scan rate of 1.66 mV/s in a potential range from corrosion potential to 500 mV to return to corrosion potential, using a potentiostat/galvanostat controlled with a personal computer. All the tests were conducted using the same running parameters and thereby the exposure time to the electrolyte was the same in every sample.

## 12.3 Results and Discussion

The optical micrograph in Fig. 12.3 shows the phases present in the base metal in the as-received condition. An austenitic microstructure (γ-phase) is observed with twin boundaries and δ-ferrite is present as small points of dark color.

Figure 12.4 shows the welding profiles obtained. Although it is appreciated full penetration for every weld, the welds presented lack of lateral fusion to different extents. The best condition is, however, when the magnetic field is applied perpendicular in one direction, Fig. 12.4a, the defect is less than 10 % of the thickness of

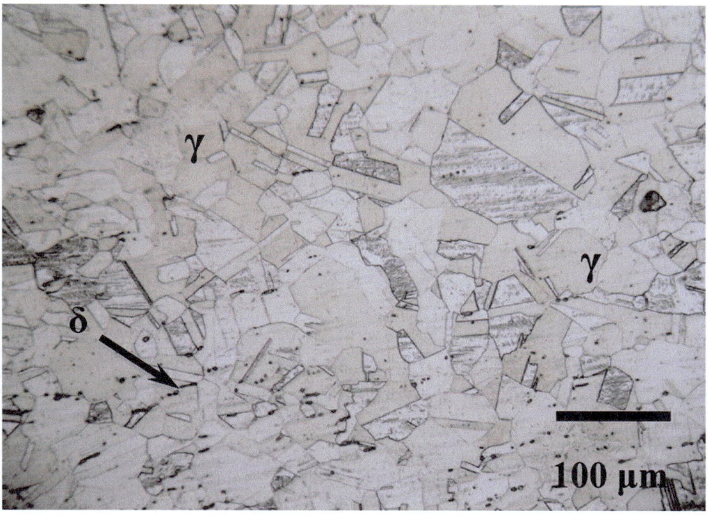


**Fig. 12.3** Microstructure of the base metal in the as-received condition

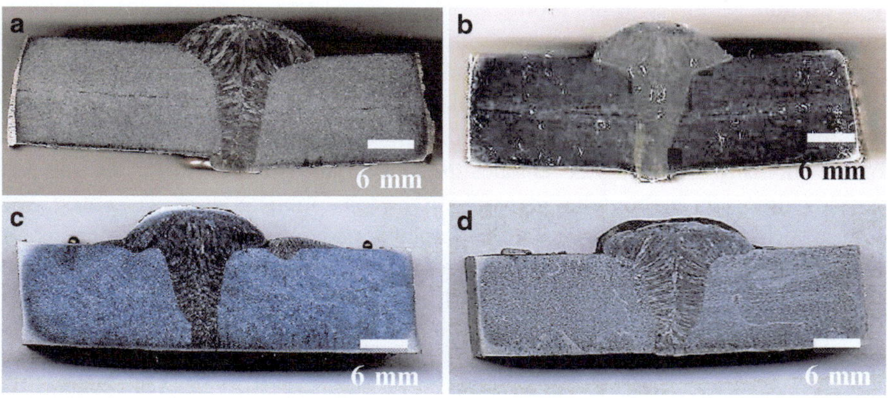


**Fig. 12.4** Macrographs of the weld profiles. (**a**) Perpendicular magnetic field in one direction. (**b**) Without magnetic field. (**c**) Perpendicular opposite axial and (**d**) Perpendicular opposite

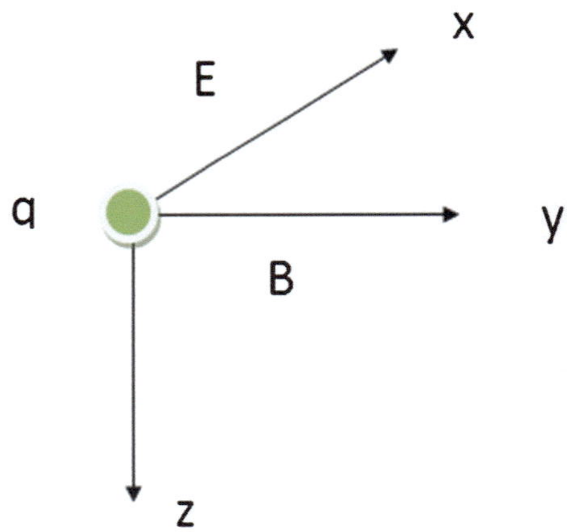


**Fig. 12.5** Coordinates system showing the application of the electric field, *E*, and magnetic field, *B*, on the electric charge, *q*, during the welding process

the parent plates. In this instance, the resultant electromagnetic field produced a major attraction electromagnetic force on the plasma of the electric arc, according to the Lorentz Force equation (12.1) and Fig. 12.5.

$$F_{\mathrm{EM}} = q\left[\vec{E} + \vec{v} + \vec{B}\right] \tag{12.1}$$

where $F_{\mathrm{EM}}$ is the electromagnetic force, $q$ is the electric charge, $\vec{E}$ the electric field, $\vec{v}$ is the electric charge rate, and $\vec{B}$ is the magnetic field. From Eq. (12.1) it is possible to obtain Eq. (12.2) (($w$) is the angular frequency), which gives the behavior of the electric arc plasma in Fig. 12.5.

$$\frac{d^3X}{dt^3} + w^2\frac{dX}{dt} = 0 \tag{12.2}$$

Figure 12.4b shows the profile of the weld without magnetic field, this weld has very similar geometry to the profile shown in Fig. 12.4a, that is because the magnetic field generated by the electric current of the weld is generated in the same direction that the external magnetic field applied, however, a close observation along the weld/BM interface revealed that there is approximately 50 % lack of lateral fusion in this weld. On this basis, the application of the external magnetic field of low intensity effectively accelerates the plasma into the joint filling the gap and melting almost throughout the sidewalls for the separation between plates used. The reduction in lack of lateral fusion of the plates occurs because the perpendicular electromagnetic field applied turns helicoidally the path of the plasma within the joint design during welding. A similar effect was reported by Hanove [10].

Figure 12.4c, d show the profiles of the welds deposited with the perpendicular magnetic field axial and perpendicular to the welding direction, respectively. In both welds, there is 25 % lack of lateral fusion at the bottom of the weld. This defect is due to the effect of the magnetic field gradient, which deviates the plasma during welding.

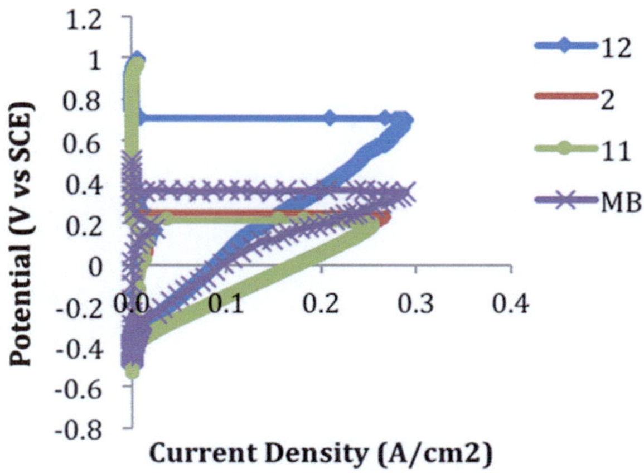


**Fig. 12.6** Typical curves obtained from the DL-EPR test

**Table 12.3** Estimation of DOS from the DL-EPR curves

| Sample | $I_a$ | $I_r$ | DOS |
|---|---|---|---|
| As-received base metal | 0.289 | 0.025 | 0.0865 |
| Weld without magnetic field (2) | 0.2877 | 0.02578 | 0.0896 |
| Weld with magnetic field (11) | 0.258 | 0.0138 | 0.0534 |
| Weld with magnetic field (12) | 0.270 | 0.0175 | 0.0646 |

Regarding the effect of the magnetic field in corrosion resistance, Fig. 12.6 presents the results of the DL-EPR tests in the HAZ as compared to the behavior of the as-received stainless steel and the weld obtained without magnetic field. From the plot, it is clear that the as-received material and the sample from the weld without magnetic field exhibit the largest $I_r/I_a$, namely 0.0865 and 0.0896 as calculated in Table 12.3, respectively. It is established that when the $I_r/I_a>0.05$ the material is sensitized. Conversely, the samples of the welds performed with opposite axial and perpendicular opposite magnetic fields were found to exhibit a lower susceptibility to intergranular corrosion. These findings reveal that the orientation of magnetic fields applied during welding may have a profound impact in the microstructural condition induced by this practice so that the susceptibility to intergranular corrosion maybe enhanced or reduced.

This result demonstrates that it is very important to know the orientation of the low intensity magnetic field when it is applied during welding, because the effect of the electromagnetic interaction generated in solid state is completely different according to the short-range diffusion of chromium in the HAZ. Curiel et al. [12] found, by transmission electron microscopy, that the electromagnetic interaction produced by the application of an axial magnetic field raised the Cr content healing Cr depleted zones in the HAZ of an austenitic 304 stainless steel.

The samples subject to the DL-EPR test were observed in the scanning electron microscope (SEM) and the features of the surfaces are shown in Fig. 12.7. The images reveal that the AISI 304 stainless steel in the as-received condition presents considerable damage at preferential sites such as pre-existent carbides. Welding in

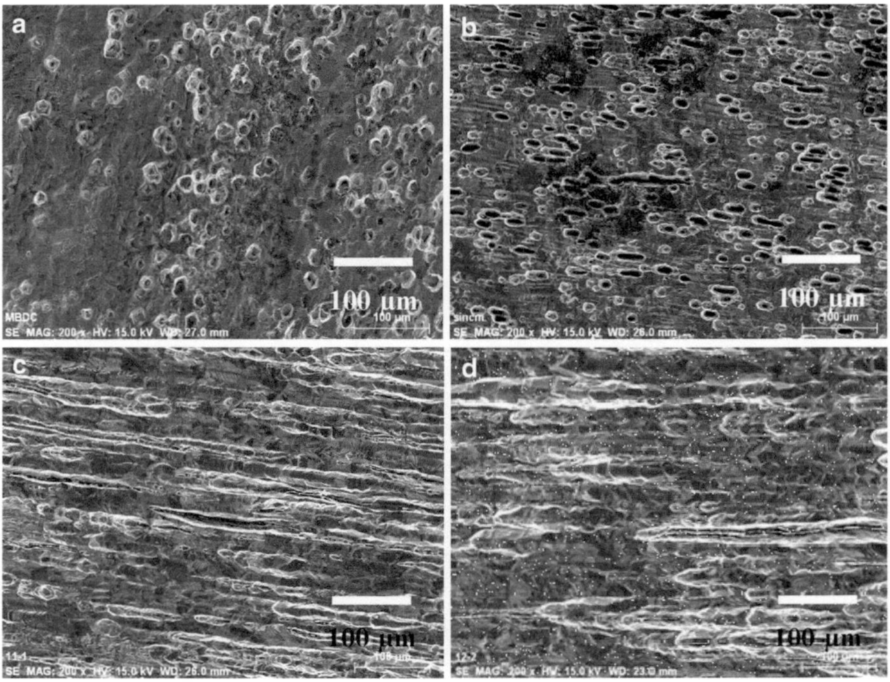


**Fig. 12.7** SEM images after the DL-EPR test. (**a**) Base metal in the as-received condition. (**b**) Weld without MF. (**c**) With axial opposite MF and (**d**) Perpendicular opposite MF

absence of magnetic field (Fig. 12.7b) makes the plates more susceptible to localized corrosion in the HAZ. When the magnetic field is applied in two forms opposite axial and perpendicular to the welding direction, the DOS value was reduced according to Table 12.3. However, Fig. 12.7c, d disclosed very different corrosion features. For these cases, corrosion occurs more in a uniform fashion rather than localized. Bear in mind that uniform corrosion is preferred over localized corrosion since the first allows prediction of residual life of a component whereas with the second this estimation is not possible and premature failure of components is highly likely. The objective of this study was not to induce uniform corrosion instead of localized corrosion. The goal was to gain understanding about the effect of applying different orientations of magnetic fields during welding in the corrosion behavior of AISI 304 stainless steel. Thus, the experimental setup previously used [11, 12] leads to the best results in terms of reducing the susceptibility to intergranular corrosion in the HAZ.

## 12.4 Conclusions

Fusion welding of AISI 304 austenitic stainless steel plates with 12.7 mm in thickness was possible using a non-conventional joint preparation in a single welding pass with and without the application of an external magnetic field.

However, major lack of lateral fusion was seen when plain welding. This defect was reduced when a perpendicular external electromagnetic field was applied with different orientations due to the acceleration of the plasma into the joint following a helicoidal path. According to the $I_r/I_a$ ratio, the as-received plates presented some DOS, which was induced during the manufacturing process. This DOS was worsened in the HAZ of the weld obtained without the application of external magnetic field. On the other hand, when the opposite axial and perpendicular magnetic fields are applied, the DOS value is reduced. However, the samples exhibited generalized corrosion instead of localized corrosion. Thus, this study reveals that the orientation of the magnetic field applied during welding may have a major role in conditioning the microstructural characteristics of the HAZ so that localized corrosion of this zone maybe improved, as seen in [12], or change into uniform corrosion as observed in this work.

**Acknowledgment** The authors thank CONACyT for financial support.

## References

1. Lin P, Palumbo G, Erb U, Aust T (1995) Influence of grain boundary character distribution on sensitization and intergranular corrosion of alloy 600. Scr Met Mater 33:1387–1392
2. Palumbo G, Aust KT (1990) Structure-dependence of intergranular corrosion in high purity nickel. Acta Metall Mater 38:2343–2352
3. Kokawa H, Shimada M, Sato M (2000) Grain-boundary structure and precipitation in sensitization austenitic stainless steel. J Mater 52:34–37
4. Lippold JC, Kotecki DJ (2005) Welding metallurgy and weldability of stain-less steels, 1st edn. Wiley, New Jersey
5. Kokawa H, Shimada M, Sato YS (2000) Grain-boundary structure and precipitation in sensitized austenitic stainless steel. JOM 52:34–38
6. Kokawa H (2005) Weld decay-resistant austenitic stainless steel by grain-boundary engineering. J Mater Sci 40:927–932
7. Kaul R, Parvathavarthini N, Ganesh P, Mulki SV, Samajdar I, Dayal RK, Kukreja LM (2009) A novel pre-weld laser surface treatment for enhanced intergranular corrosion resistance of austenitic stainless steel weldments. Weld J 32:233–242
8. Watanabe Y, Takeda T, Sato H (2006) Effect of magnetic field on weld zone by spot-welding in stainless steel. ISIJ 46:1292–1296
9. Fujiii H, Tsurekawa S (2011) Diffusion of carbon in iron under magnetic fields. Phys Rev B 83:405–412
10. Hanove E (1991) Magnetic arc blow. Weld Int 5:995–999
11. Curiel FF, García R, López VH, González-Sánchez J (2011) Effect of magnetic field applied during gas metal arc welding on the resistance to localised corrosion of the heat affected zone in AISI 304 stainless steel. Corros Sci 53:2393–2399
12. Curiel FF, García R, López VH, García MA, Lemus J (2013) Transmission electron microscopy in the heat affected zone of an AISI 304 austenitic stainless steel welded with the application of a magnetic field of low intensity. Mater Trans 1:122–125
13. Majidi AP, Streicher MA (1984) The double loop reactivation method for detecting sensitization in AISI 304 stainless steel. Corrosion 40:584–593

# Chapter 13
# On the Effect of Crosshead Velocity on Polycrystalline Flow During Tension Testing of a 430 Stainless Steel

**Elizabeth Garfias-García, Juan Daniel Muñoz-Andrade, Pablo Gerardo Rodríguez-López, Miriam Aguilar-Sánchez, and Benjamín Vargas-Arista**

**Abstract** A systematic study of polycrystalline flow during in situ tension testing of 430 stainless steel was done at 25 °C using a tensile module coupled in a field emission scanning electron microscope at three different crosshead velocities. By mapping the polycrystalline flow it was found that the trajectory of the crystals during an irreversible deformation process followed by a hyperbolic motion. Such behavior is related to the microstructural evolution during tensile testing, as was observed, the polycrystalline irreversible deformations are promoted by coupled mechanisms: dislocation dynamics at inner individual crystals and granular group movement, named cellular dislocations. There are some important differences in the microstructure evolution and ductility related to the effect of the crosshead velocity. In connection, with the ductile behavior of SEPCS, the fractography study exhibited a notable change in the density and diameter of microvoids, which were changed from large to small in size as a crosshead velocity increased.

**Keywords** Mapping polycrystalline flow • Active energy • AISI 430 • FESEM • Fractography

---

E. Garfias-García (✉) • J.D. Muñoz-Andrade • P.G. Rodríguez-López • M. Aguilar-Sánchez
División de Ciencias Básicas e Ingeniería, Departamento de Materiales, Universidad Autónoma Metropolitana-Azcapotzalco, UAM-Azc, Av. San Pablo 180, Col. Reynosa, 02200 D. F. México, México
e-mail: egg_liz@hotmail.com; elgg@correo.azc.uam.mx; jdma@correo.azc.uam.mx; metallmorph@hotmail.com; mas@correo.azc.uam.mx

B. Vargas-Arista
División de Estudios de Posgrado e Investigación, Instituto Tecnológico de Tlalnepantla, Av. Instituto Tecnológico s/n, Col. La Comunidad, 54070 Tlalnepantla de Baz, Edo. de México, México
e-mail: benjaminvar@yahoo.com.mx



R. Pérez Campos et al. (eds.), *Materials Characterization*,
DOI 10.1007/978-3-319-15204-2_13

## 13.1 Introduction

AISI 430 stainless steel is susceptible to dynamic recrystallization. This has led to grain refinement during plastic deformation to produce a better ductility of the metal through dynamic recrystallization. This phenomenon is associated with dislocation density and low stacking fault energy. It is also remarkable that during the forming of metals, the higher strain rate and lower temperature in thermomechanical working promotes an increased dislocation density and grain refinement due to recrystallization. This AISI 430 steel has a great interest, due to their technological applications in engineering and widely used in metalworking process [1], even though there is less information associated with the investigation of the mechanisms of irreversible deformation at different tension test velocity conditions that contribute to the ductility of metalworking processes in this steel.

In recent studies, it has been observed that the polycrystalline flow in austenitic AISI 304 stainless steel obeys a hyperbolic trajectory during the process of irreversible deformation under unidirectional stress with a constant crosshead velocity. This behavior occurs similarly to the general trend of the flow of matter in an oil-emulsion as suggested by Ashby and Verrall [2], who suggested that granular flow maps are one way of identifying a non-uniform flow. A physical theory has been developed pertaining to polycrystalline flow associated with the determination of the activation of energy for plastic flow which establishes the Hubble flow unification and the plastic flow of polycrystalline systems to spatially widespread, commonly called metallic materials [3]. Recently, studies were performed on polycrystalline metal flow, such as materials like AISI austenitic stainless steels [4] and in a Pb-50%Sn alloy at room temperature where it was possible to calculate the activation of energy associated with the deformation process during the unidirectional tension test [5]. Therefore, the present work characterizes the mechanical behavior at room temperature of AISI 430 sheet steel using the mapping of the polycrystalline flow (MPF), during axial tension test of spatially extended polycrystalline systems (SEPCS) which allows determining relevant aspects of the phenomenology and mechanics of polycrystalline metals flow and alloys during irreversible deformation.

## 13.2 Experimental

The material used in this work was a sheet of AISI 430 stainless steel and all the samples were obtained from the cross section of the direction of rolling. A metallographic preparation was prepared before the tension test on the surface of the tensile test specimen. By applying the new technique described in the references [4–7], the coordinate reference system was inscribed on the polished and etched surface of the center of the specimen by using a pyramidal shaped micro Vickers indenter and a diamond pyramidal set of indentations were used to follow the experiment performed by field emission scanning electron microscopy (FESEM). Figure 13.1

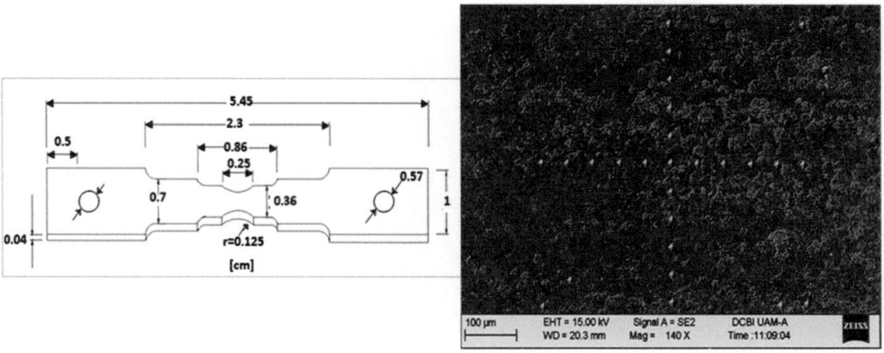


**Fig. 13.1** The initial dimensions of the tension test sample and the coordinate system inscribed on the surface of the center of a tension test sample of 430 stainless steel

shows the initial dimensions of the tension test sample and the coordinate system inscribed on the surface of a center of a tension test sample. Axial tensile testing of AISI 430 was performed on a special Kammarth & Weiss Company module, coupled in the FESEM with a capacity of 500 N. The experiment was performed at different crosshead velocities of 0.19, 0.38, and 0.57 mm/min, respectively.

During different steps of the irreversible deformation process the tension test was stopped in order to take microstructure evidence of the evolution of SEPCS at different magnifications; all experiments were conducted up to the fracture. The microstructures of SEPCS taken in the FESEM with a secondary detector were taken at 400× and at a mark of 20 µm. The microstructures for the fractographic analysis were taken for all samples, with a secondary detector, at different magnifications.

## 13.3 Results and Discussion

The irreversible deformation process was performed on the samples, at different crosshead velocities of 0.19 mm/min with six steps of deformation, 0.38 mm/min with eight steps of deformation and 0.57 mm/min with four steps of deformation; all the tension tests were conducted up to the fracture, following a displacement of 12 grains. Figure 13.2a shows the original quadrant and Fig. 13.2b shows the last steps during deformation at a crosshead velocity of 0.38 mm/min, and also shows the reference system for the 12 grains.

### *13.3.1 Mapping Polycrystalline Flow*

During tensile testing, the evolution of the microstructure of 430 stainless steel was achieved by the FESEM study of surfaces of the tensile specimens under three different values of crosshead velocities. Figure 13.3 shows the trajectory of

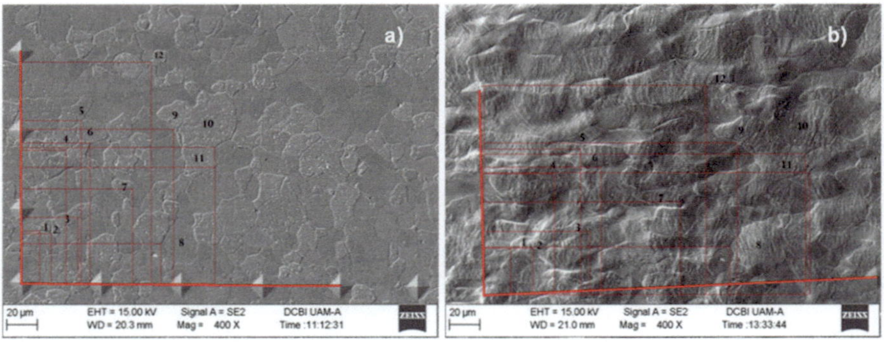


**Fig. 13.2** Initial (**a**) and last (**b**) reference systems for a crosshead velocity of 0.38 mm/min

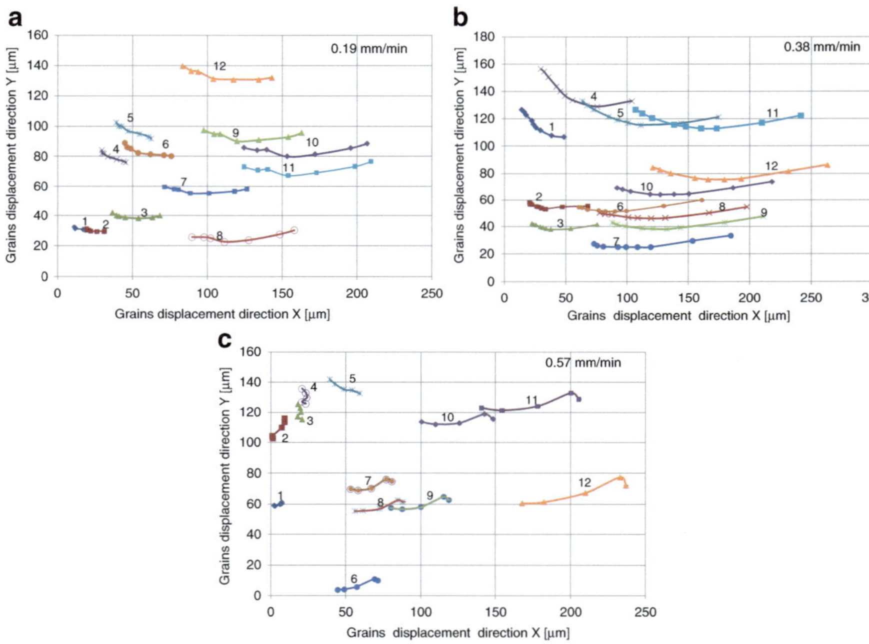


**Fig. 13.3** MPF curves of 430 stainless steel obtained under various crosshead velocity conditions: (**a**) 0.19 mm/min, (**b**) 0.38 mm/min, and (**c**) 0.57 mm/min

polycrystalline flow obtained under various constant crosshead velocity conditions, in which the general characteristics of the MPF curves are similar at all velocity conditions.

By comparing the MPF graphs, we can see that at a crosshead velocity of 0.38 mm/min the grain flow is more extended than at the higher velocities, 0.57 mm/min. This behavior is due to variation of the strain rate during tension test at constant crosshead velocity. In general at low crosshead velocity the range of strain

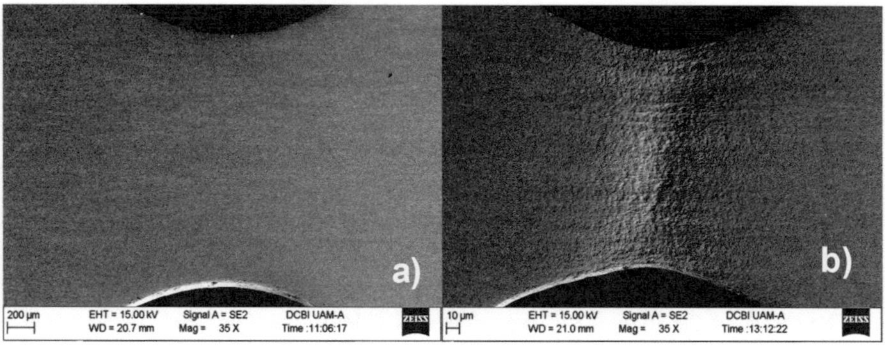


**Fig. 13.4** Surface of the center of a tension test sample of 430 stainless steel at (**a**) initial condition and (**b**) with an irreversible deformation of $\varepsilon = 0.31$ associated with crosshead velocity of 0.38 mm/min

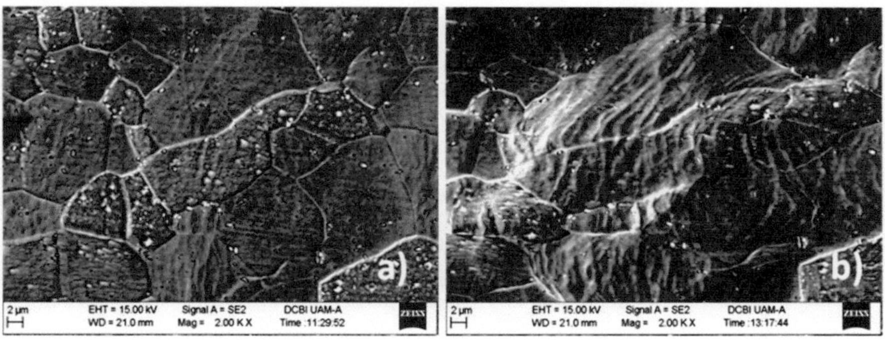


**Fig. 13.5** Surface microstructure of the center of a tension test sample of 430 stainless steel at (**a**) initial condition and (**b**) the microstructure evolution with an irreversible deformation $\varepsilon = 0.31$ associated with crosshead velocity of 0.38 mm/min

rate is lower than at high crosshead velocity. Recently, it was shown that strain rate associated with dynamics of mobile dislocation plays an essential role because the activation energy for plastic flow increased as irreversible deformation process is happening. In consequence, the activation energy for plastic flow decreased as strain rate is increased and the ductility of SEPCS is controlled in the same manner [8].

Figure 13.4a represents the initial surface conditions of the tensile test sample of 430 stainless steel and Fig. 13.4b the same surface with an irreversible deformation ($\varepsilon = 0.30$) obtained under crosshead velocity of 0.38 mm/min, where the trajectory of polycrystalline flow during irreversible deformation is related to the microstructure evolution.

It can be concluded from Fig. 13.5b that total strain ($\varepsilon = 0.30$) is promoted by coupled mechanisms like dislocation dynamics (DD) at inner individual crystals, with a strain role of $\varepsilon_{DD} = 0.15$ and cellular dislocations dynamics (CDD), with a strain part of $\varepsilon_{CDD} = 0.15$, formed at the surface of the tensile sample [9–13].

### 13.3.2 Fractography

Additional indirect measurement of the effect of crosshead velocity on polycrystalline flow, during tensile test of 430 stainless steel was achieved by the scanning electron microscopy study of surfaces obtained from tensile fractured specimens under different crosshead velocities of 0.19, 0.38, and 0.57 mm/min, respectively. The fracture surfaces exhibited a microvoids morphology characteristic of ductile type of failure associated with the mechanism of nucleation, growth and coalescence of microcavities, as can be seen in Fig. 13.6.

The effect of the increase in the crosshead velocity on fractured surfaces was a notable change in density and diameter of microvoids, which was associated with the reduction of polycrystalline plastic deformation. Appreciable reduction in voids diameter and increase in their density across the fracture surface were observed as a consequence of the increase in crosshead velocity related to the minor growth of microcavities resulting in their less coalescence. Moreover, the morphology of microvoids changed from large in size (average size of 11.56 μm) to medium in size (average size of 10.29 μm) and small in size (average size of 6.92 μm), correspondingly, with the increase in testing crosshead velocity. Additionally, the cross-sectional area of the fractured surfaces was decreased with the increase in crosshead velocity.

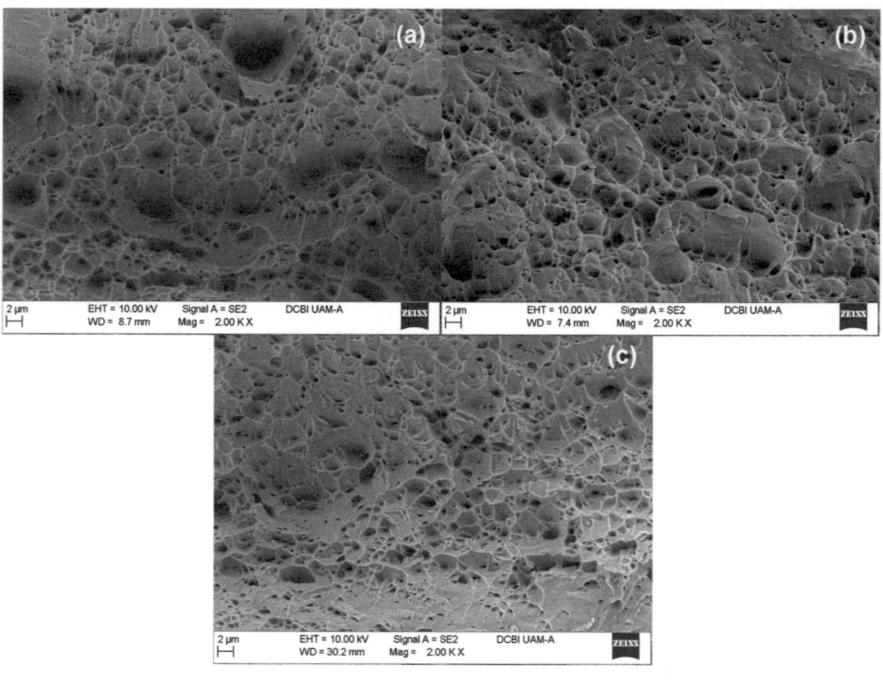


**Fig. 13.6** Fractured surface morphology characteristic of the ductile type failure showing microvoids with different density and diameter: (**a**) large voids at velocity of 0.19 mm/min, (**b**) medium voids at 0.38 mm/min, and (**c**) small voids at 0.57 mm/min

At 0.19 mm/min of crosshead velocity, the fracture surface showed voids with large and medium in size on uniform cross-sectional fracture area (Fig. 13.6a) while velocity of 0.38 mm/min produced medium and small voids on a reduced area of fracture surface (Fig. 13.6b). Finally, the velocity of 0.57 mm/min generated a fracture surface of discontinuous area containing small and shallow voids, see Fig. 13.6c).

## 13.4 Conclusions

The systematic study of polycrystalline flow obtained in this work allows the describing of some qualitative aspects related to the phenomenology and mechanics of polycrystalline flow in SEPCS of 430 Stainless Steel at different crosshead velocities of 0.19, 0.38, and 0.57 mm/min and the main results can be summarized as follows: The mapping of the polycrystalline flow in SEPCS establishes that it is related with the hyperbolic motion for all crosshead velocities, where the total strain is promoted by coupled mechanisms like dislocation dynamics at inner individual crystals and cellular dislocations dynamics formed on the surface of SEPCS. Another important point to be noted is that the ductility of SEPCS is controlled with the variation of the strain rate during tension test at constant crosshead velocity. On the other hand, in connection with the ductile behavior of SEPCS, the fractography study exhibited a notable change in the density and diameter of microvoids, which were changed from large to small in size as a crosshead velocity increased.

**Acknowledgments** The authors thank Laboratorio de Microscopia Electrónica de la DCBI UAM Azcapotzalco unit, for the use of their installations and the use of the special tension module coupled with the FESEM. EGG, MAS, and BVA gratefully acknowledge the SNI for the distinction of their membership and the stipend received.

## References

1. Kim SI, Yoo Y-Ch (2002) Continuous dynamic recrystallization of AISI 430 ferritic stainless steel. Met Mater Int 8:7–13
2. Ashby MF, Verrall RA (1973) Diffusion-accommodated flow and superplasticity. Acta Metall 21(2):149–163
3. Muñoz-Andrade JD (2007) On the hyperbolic flow manifested during the irreversible deformation processes in spatially extended crystalline systems. AIP Conf Proc 907:1283–1288
4. Muñoz-Andrade JD (2013) Systematic study of polycrystalline flow during tension test of sheet 304 austenitic stainless steel at room temperature. AIP Conf Proc 1567:796–799
5. López-Ramírez AJ, Muñoz-Andrade JD, Garfias-García E, Aguilar-Sánchez M (2014) Systematic study of polycrystalline flow during in situ tension test by SEM of Pb-50%Sn at room temperature. Mater Sci Forum 793:113–118
6. Muñoz-Andrade JD, Mendoza-Allende A, Torres-Villaseñor G, Montemayor-Alderete JA (2001) Mapping of the granular flow during superplastic deformation of microsample of Zn-20.2%AI-1.8%Cu alloy at room temperature. J Mater Sci 36:795–799

7. Muñoz-Andrade JD (2007) Mapping of the granular flow during the irreversible deformation processes in spatially extended polycrystalline systems. Mater Sci Forum 561–565:901–904
8. Muñoz-Andrade JD (2008) The activation energy for plastic flow in spatially extended polycrystalline systems during tension test. Int J Mater Form 1:81–84
9. Morral JE, Ashby MF (1974) Dislocated cellular structures. Acta Metall 22:567–575
10. Zelin MG, Dunlap MR, Rosen R, Mukherjee AK (1993) The direct observation of cooperative grainboundary sliding and migration during superplastic deformation of leadtin eutectic in shear. J Appl Phys 74:4972–4982
11. Zelin MG, Mukherjee AK (1995) Cooperative phenomena at grain boundaries during superplastic flow. Acta Metall Mater 43:2359–2372
12. Zelin MG, Mukherjee AK (1996) Geometrical aspects of superplastic flow. Mater Sci Eng A 208:210–225
13. Zelin M, Guillard S, Mukherjee A (2001) Cellular dislocations: experiment and modelling. Mater Sci Eng A 309–310:514–519

# Part V
# Characterization of Composite Materials

This part includes two chapters related to the characterization of metal matrix composites (MMC). These kinds of materials are of interest in automotive, aerospatial, and electronic industry mainly. MMC exhibits excellent mechanical properties such as wear and abrasion resistance, high strength, chemical stability, and dimensional stability to high temperatures. These MMCs have been displacing to the commercial steels, mainly when a specific property is needed.

One of the chapters presents results about microstructural and mechanical characterization of Al/SiC composite joined by friction stir welding (FSW) technique, which appears as the most promising joining process compared with conventional fusion welding processes, mainly for the ability to join without compromising their ceramic reinforcements. This study was focused to evaluate the microstructural changes using diverse parameters in welding process.

The second chapter studied the corrosion process of Al–Cu/TiC composite compared with the corrosion of Al–Cu alloy immersed in synthetic seawater. Al–Cu/TiC composite was synthesized to satisfy specify mechanical properties and in order to generate information about the corrosion kinetic of this composite, synthetic seawater was used as corrosive environment.

The main techniques used in the characterization of these composites are:

- Scanning electron microscopy (SEM).
- Optical microscopy (OM).
- Mechanical tests.
- Electrochemical test (polarization curves).

# Chapter 14
# Effect of FSW Parameters on Microstructure of Aluminum Matrix Composites Joints

**O. Cuevas Mata, A.F. Miranda Pérez, F.J. García Vázquez, G.Y. Pérez Medina, and F.A. Reyes Valdés**

**Abstract** Automotive and aerospace industries are maintaining constant development and innovation, demanding for materials that can fulfill emissions, fuel efficiency, and security regulations. Recently, the interest in automotive applications of metal matrix composites (MMC) is increasing. MMC exhibits excellent mechanical properties such as wear and abrasion resistance, high strength, chemical stability, and dimensional stability to high temperatures. In order to join MMC, friction stir welding (FSW) appears as the most promising joining process compared with conventional fusion welding processes, mainly for the ability to join without compromising their ceramic reinforcements. The aim of this study was to evaluate the microstructural changes of diverse parameters of aluminum matrix reinforced with 20 % wt. Al/SiC composite was welded with different process parameters such as tool rotation speed and axial force using FSW. In order to determine the different microstructural zones characterization of the samples was carried out by optical microscopy (OM) and scanning electron microscopy (SEM). In addition, an evaluation of mechanical property such as microhardness of the different zones of the joint in the aluminum matrix composite samples was performed.

**Keywords** Microstructure • Aluminum • Joint • Friction stir welding (FSW) • Metal matrix composites (MMC)

## 14.1 Introduction

The new competition for enhanced quality products in the automotive and aerospace sectors requires to make improvements and implementations in industrial processes for the product to be accepted in a competitive market. Different materials in

O. Cuevas Mata (✉) • A.F. Miranda Pérez • F.J. García Vázquez • G.Y. Pérez Medina
F.A. Reyes Valdés
Corporación Mexicana de Investigación en Materiales,
Ciencia y Tecnología No. 790 Col. Saltillo 2000, Saltillo, Coahuila, México
e-mail: octavio.cuevas@comimsa.com; argelia.miranda@comimsa.com; felipegarcia@comimsa.com

 
R. Pérez Campos et al. (eds.), *Materials Characterization*,
DOI 10.1007/978-3-319-15204-2_14

automobile or aerospace vehicles are subject to different working conditions and must have specific qualities such as weight reduction with a development in impact resistance, reduction of emissions, improved comfort and security. Therefore, the use of aluminum alloys has recently increased due to the different characteristics that make them attractive for automotive applications such as low density, good resistance to corrosion, low thermal expansion, and established casting techniques for mass production [1]. Aluminum matrix composites (AMCs) have emerged in the last years as materials formed by an aluminum matrix reinforced with a particular form of reinforcement. Reinforcements can be embedded as either continuous or discontinuous; in short, long fibers or particles for discontinuous. This material has applications in sports equipment, automotive and aerospace industries due to its improved features. Those materials are joined using fusion welding as traditional joining process, which often leads to deterioration of the AMCs obtaining as a result reduced joint strength due to solidification shrinkage, oxide inclusions, porosity, distortion, increasing of residual stress, formation of intermetallic compounds in the case of chemical reactions between matrix and reinforcement [2]. Problems previously mentioned can be eliminated if AMCs are welded by solid state welding process [3–6].

Friction stir welding (FSW) was invented by The Welding Institute (TWI), UK in 1991, as a low heat input solid state welding [7]. In the FSW process, a non-consumable hard rotating tool with a specially designed probe travels down the length of contacting metal plates, and it produces highly plastically deformed zone through the associated stirring action and consolidates the retreating side by the axial force acting through the tool shoulder [2].

The friction between the tool shoulder and the plate top surface produces the thermo-mechanically affected zone, also produces the plastic deformation of the material in contact with the tool [8]. In the case of the material, it does not reach the melting point obtaining good results in higher mechanical properties of the deformed material compared to the parent material [9–11].

For AMCs very hard tool materials with special geometries are needed to produce the joints because of the very different mechanical properties of the matrix and the ceramic reinforcement [12].

In FSW, the different parameters such as tool rotational speed, welding speed, axial force, tool geometry, tool material, tool tilt angle, clamping force, and geometry have important effect in AMC, but the first four parameters play a significant role to produce sound weld joints [13].

In automotive and aerospace industries, FSW process has many advantages, including mechanical properties (tensile and fatigue), improved process robustness, lack of consumables, less health and environmental issues, and operating advantages [14].

The aim of this work is to study the effect of FSW parameters on the microstructure of welds in the AMC A359/SiC/$20_p$ sheet, which is manufactured by squeeze casting infiltration.

Therefore, the metallographic characterization is very important for the effect on the properties due to stirred zone that may occur in this type of material, when is subjected to a FSW process.

## 14.2 Experimental

The AMC employed for this experimentation is made with 20 % volume percentage reinforcing SiC particles, distributed in an A359 alloy matrix. The AMC was obtained through the purchase with the company MC21 Incorporated, USA. The samples were cut into small plates: 62 mm × 22 mm × 5 mm thickness in size. The sample will be referred to as the Al/SiC composite, throughout this study. The nominal composition of the A359/SiC/$20_p$ is shown in Table 14.1 and the mechanical properties in Table 14.2.

Al/SiC composites were manufactured by squeeze casting infiltration and afterwards hot rolled in order to obtain sheets.

To develop the experiment, the input variables had to be determined: tool rotation speed (rpm), welding speed (mm/min), and axial force (kN), continuing with three welds on the plates which were performed on an FSW machine using a tool made of high strength steel-high temperature having a shoulder diameter of 20 mm and a pin diameter of 8 mm, the parameters applied to weld the Al/SiC composite are shown in Table 14.3.

The Al/SiC composite joining was conducted in Helmholtz-Zentrum Geesthacht Solid State Joining Processes Department, with FSW equipment. In addition, optical microscopy (OM) and scanning electron microscopy (SEM) were employed in characterizing the exhibited microstructures of the welded plates of Al/SiC composite. The sample surfaces were etched with Keller reagent for 15 s to facilitate the observation of the AMC metallographic structures. The etching solution is showed in Table 14.4, this reagent allows to identify the matrix alloy and SiC particles. Moreover, microhardness test was performed in order to evidence the variations along the joint of the Al/SiC composite.

**Table 14.1** Chemical composition of the A359/SiC/$20_p$ AMC

| Element | Si | Fe | Cu | Mn | Mg | Zn | Ti | Al |
|---|---|---|---|---|---|---|---|---|
| Percentage | 8.5–9.5 | 0.2 | 0.2 | 0.1 | 0.50–0.70 | 0.1 | 0.2 | 89–91 |

**Table 14.2** A359/SiC/$20_p$ mechanical properties

| SiC (% vol) | Density (g/cm$^3$) | Tensile strength (MPa) | Yield strength (MPa) | % Elong. | Young's modulus (GPa) | Thermal conductivity (W/mK) |
|---|---|---|---|---|---|---|
| 20 | 2.72 | 175 | 160 | 5 | 100 | 150 |

**Table 14.3** FSW parameters

| Weld | Tool rotation speed (rpm) | Welding speed (mm/min) | Axial force (kN) |
|---|---|---|---|
| 1 | 1,000 | 6 | 8 |
| 2 | 1,000 | 6 | 10 |
| 3 | 1,200 | 6 | 10 |

**Table 14.4** Keller composition

| Composition | Symbol | Volume (ml) | Volume (%) |
|---|---|---|---|
| Hydrofluoric acid | HF | 1 | 6.90 |
| Hydrochloric acid | HCl | 1.5 | 10.34 |
| Nitric acid | $HNO_3$ | 2.5 | 17.24 |
| Distilled water | $H_2O$ | 9.5 | 65.52 |

**Fig. 14.1** Optical macrograph of sample 2 Al/SiC composite joined by friction stir welding

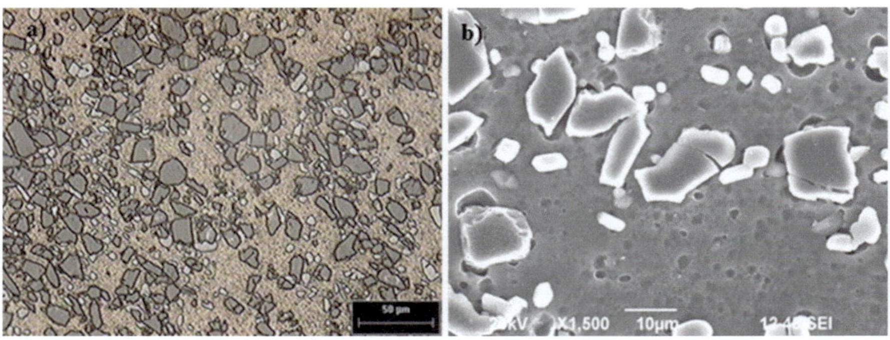


**Fig. 14.2** Homogeneous distribution of particles in Al/SiC composite. (**a**) Optical microscopy (**b**) Scanning electron microscopy

## 14.3 Results and Discussion

The Al/SiC composite under this investigation was successfully joined by FSW. Figure 14.1 shows the macrostructure of the sample 2 with the advancing and retreating side. No gross defects such as voids or tunnel defect are detected and a homogeneous nugget is visible in the joint.

Figure 14.2 evidences the microstructures of the Al/SiC composite as received condition. The composite microstructure exhibits a suitable SiC particle distribution in the matrix aluminum, concentrated in the interdendritic regions, resulting in a structure called hard grain. Matrix aluminum alloy is formed by a matrix of α-phase.

Optical microscopy and SEM characterization were performed on the weld joints to reveal the directionally recrystallized zone (DXZ) in the microstructure, caused

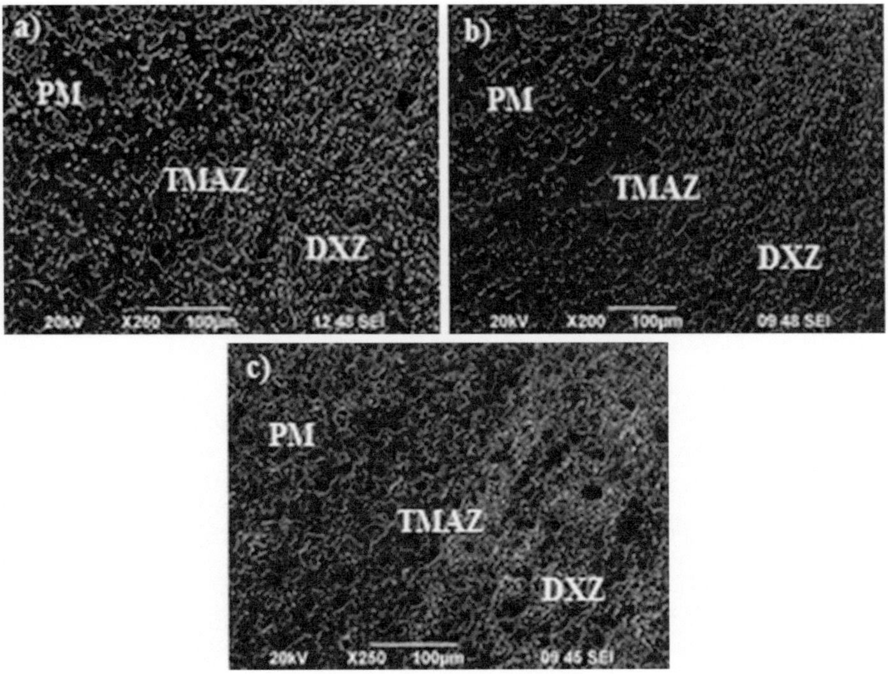


**Fig. 14.3** Micrographs obtained by SEM showing friction-stir welds made with different welding parameters on Al/SiC composite. (**a**) Sample 1, (**b**) Sample 2, and (**c**) Sample 3 shows the parent metal (PM), thermo-mechanically affected zone (TMAZ) and directionally recrystallized zone (DXZ)

by the flow of material during the joining process; DXZ is showed in Fig. 14.3. The thermomechanical affected zone (TMAZ) was detected by microstructural observations due to different particle behavior in distinct zones of the weld. In the AMC studied, particle fracture was observed in the DXZ compared with the parent material and TMAZ.

The particle rupture in this kind of joints provides a positive effect, since smaller particles stimulate the recrystallization acting as new grain nucleation sites. It leads to the formation of finer grain structure more resistant at room temperature (Fig. 14.4) [15]. SiC particles would be finer and more granular due to the solid collisions at lower temperature being a result of the more difficult plastic deformation [16]. Also SiC particles mean size were decreased and their number were increased in the stir zone which is consistent with another research [17].

It is known that the rotation speed has the same influence as the welding speed in plastic flow rate and heat input.

In this research, the particle rupture decreases from the center of the welds through the TMAZ and also into the parent material. There is no fracture particle in the parent material and also the microstructure is different compared to the DXZ in the three samples.

SEM images of the stir zone obtained at different tool rotation speed and axial force but keeping a constant welding speed of 6 mm/min are presented in

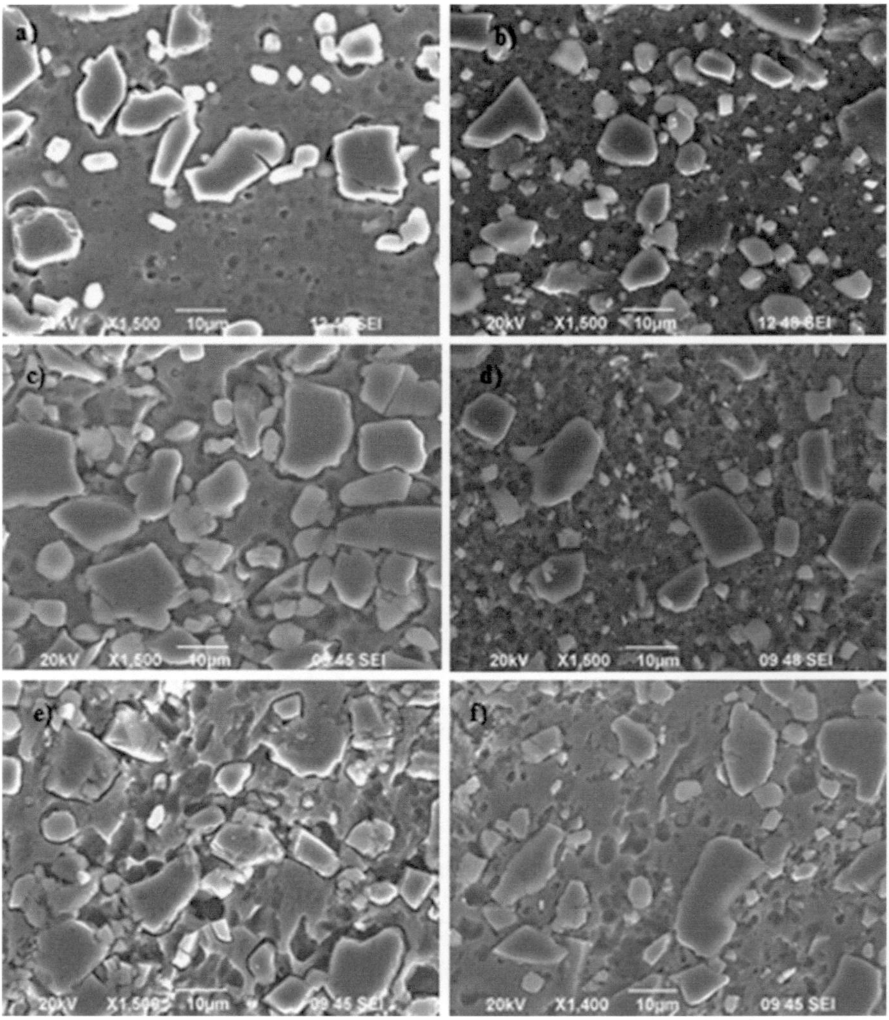


**Fig. 14.4** Differences in particles size are observed between parent material and DXZ, respectively, in Sample 1 (**a**) and (**b**), Sample 2 (**c**) and (**d**), and Sample 3 (**e**) and (**f**)

Fig. 14.4b, d, f. At any tool rotation speed, the size of de SiC particles in the DXZ was smaller than in the other zones. For all rotation speeds, the average size of the SiC particles was almost constant in the DXZ of the three samples.

Microhardness evaluation is presented in Fig. 14.5 performed along the center of the cross-section of the welds. According to the microstructure sample 1 which has the lowest axial force resulted in higher hardness value for the weld nugget (WN). In general the nugget zone of the three samples reached the highest values of microhardness. DXZ presented low values of microhardness due to the recrystallization of the SiC particles, finer and more granular due to the solid collisions at lower temperature and due to plastic deformation generated by the FSW process.

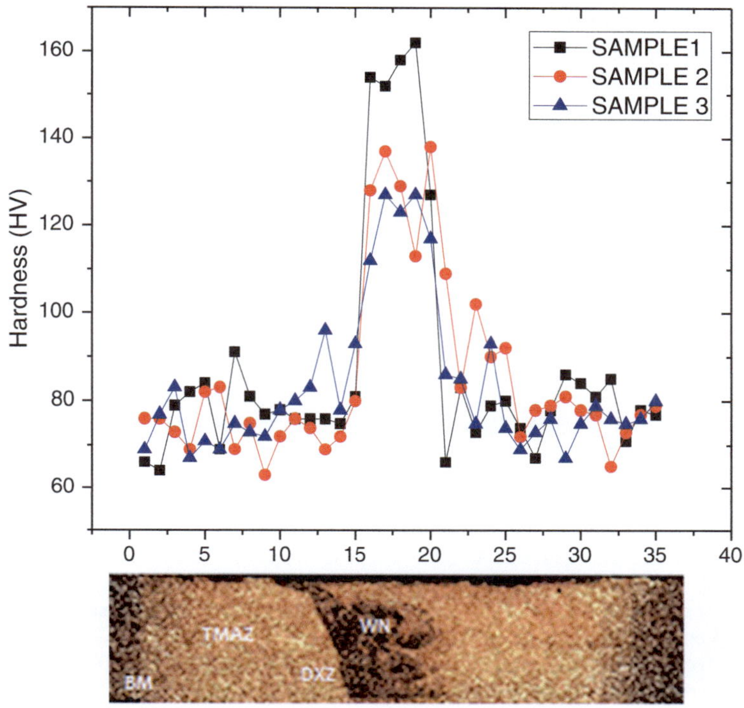


**Fig. 14.5** Hardness profile across the Al/SiC composite

## 14.4 Conclusions

An Al/SiC/$20_p$ AMC was joined by FSW under various conditions. The effect of welding parameters on the SiC particle distribution in the stir zone of the FSW joints was investigated. As a result, the following conclusions were achieved:

- FSW appeared substantially defect free and were microstructurally characterized by optical and scanning electron microscopic techniques.
- In the three samples, the DXZ are shown as fine recrystallized grains without a dendritic structure and the SiC particles are uniformly dispersed in the stir zone, the number of finer SiC particles increases during the FSW in stir zone, the size of SiC particles is smaller in DXZ than in all zones, the SiC particle size decreases when the tool rotation speed is increased.
- In the DXZ of the first and second welded samples are free of defects, but in the third one is shown a crack due to the rotation speed is greater than the other ones. The SiC particles show fragile fracture in all samples.
- The parameters have influence on the frictional heat at abutting surfaces of the samples that affects the plastic flow and consolidation of material. It has to be considered that an optimal heat must be generated at the weld zone to get a defect free weld joint.

- Microhardness always increased in the weld nugget for the three samples, and was from the base metal to WN. This behavior is primarily due to refinement of the aluminum matrix and the rupture of the SiC particles which are distributed more homogeneously in this zone.

**Acknowledgments** This study was supported by Corporación Mexicana de Investigación en Materiales (COMIMSA) and, the FSW process was performed by Helmholtz-Zentrum Geesthacht—Solid State Joining Processes Department.

## References

1. Prasad SV, Asthana R (2008) Aluminum metal-matrix composites for automotive applications: tribological considerations. Arch Civil Mech Eng 8:103
2. Storjohann D, Barabash OM, Babu SS, David SA, Sklad PS, Bloom EE (2005) Fusion and friction stir welding of aluminum-metal-matrix composites. Metall Mater Trans 36A:3237–3247
3. Ashok B, Murugan N (2014) Optimization of friction stir welding process parameters to maximize tensile strength of stir cast AA6061–T6/AlN$_p$ composite. Materials and Design 57 (2014) 383–393
4. Gui M et al (2000) 6061Al/Al-SiCp bi-layer composites produced by plasma-spraying process. Mater Lett 46:296–302
5. Torres B et al (2009) Properties and microstructure of Al-11Si/SiCp composite coatings fabricated by thermal spray. Surf Coat Technol 203(14):1947–1955
6. Pantelis D et al (1995) Formation of wear resistant Al–SiC surface composite by laser meeting injection process. Mater Sci Technol 11:299
7. Thomas WM, Nicholas ED, Needham JC, Murch MG, Temple-Smith P, Dawes CJ (1991) Friction-stir butt welding, GB Patent No. 9125978.8, International patent application No. PCT/GB92/02203
8. Sato YS, Urata M, Kokawa H, Ikeda K (2003) Hall–Petch relationship in friction stir welds of equal channel angular-pressed aluminium alloys. Mater Sci Eng A354:298–305
9. Berbon PB, Bingel WH, Mishra RS, Bampton CC, Mahoney MW (2001) Friction stir processing: a tool to homogenize nanocomposites aluminum alloys. Scr Mater 44:61–66
10. Thomas WM, Nicholas ED, Needham JC, Murch MG, Templesmith P, Dawes CJ (1992) Improvements relating to friction welding. European patent EP 0 615 480 Bl
11. Thomas WM, Johnson KI, Wiesner CS (2003) Friction stir welding-recent developments in tool and process technologies. Adv Eng Mater 5(7):485–490
12. Cavaliere P, Cerri E, Marzoli L, Dos Santos J (2004) Friction stir welding of ceramic particle reinforced aluminium based metal matrix composites. Appl Compos Mater 11:247–258
13. Karthikeyan L, Senthilkumar VS, Padmanabhan KA (2010) On the role of process variables in the friction stir processing of cast aluminium A319 alloy. Mater 31:761–771
14. Dawes CJ, Thomas WM (1996) Friction stir process welds aluminum alloys. Weld J 75:41–45
15. Humphreys FJ, Hatherly M (1995) Recrystallization and related annealing phenomena. Elsevier Science Ltd., Oxford, Hardbound
16. Kim YG, Fujji H, Tsumara T, Komazaki T, Nakata K (2006) Effect of welding parameters on microstructure in the stir zone of FSW joints of aluminum die casting alloy. Mater Lett 60:3830–3837
17. Fernandez GJ, Murr LE (2004) Characterization of tool wear and weld optimization in the friction stir welding of cast aluminum A359+20%SiC metal matrix composite. J Mater Char 52:65–75

# Chapter 15
# Electrochemical Characterization of the Aluminum–Copper Composite Material Reinforced with Titanium Carbide Immersed in Seawater

**N. Alvarez-Lemus, C.A. Leon, A. Contreras, R. Orozco-Cruz, and R. Galvan-Martinez**

**Abstract** This work shows the electrochemical study of Al–Cu/TiC composite and Al–Cu alloy immersed in synthetic seawater. Polarization curves (PCs) and the corrosion potential as a function of the time were the electrochemical techniques used to characterize the corrosion process. A typical three-electrode electrochemical cell was used, where the working electrodes (WE) were made from Al–Cu alloy and composite samples, the reference electrode was the saturated calomel electrode (SCE) and a sintered graphite rod was used as auxiliary electrode. The electrochemical measures were carried out at atmospheric pressure and room temperature, and the total exposure time was 24 h. In addition, in order to analyze the corrosion form, a superficial analysis using scanning electron microscopy (SEM) was carried out. The electrochemical results showed that the highest corrosion rate corresponding to composite samples and the global corrosion process is a mix process, that is to say, the charge transfer resistance is limited by diffusional resistance. Pitting corrosion type was observed.

**Keywords** Electrochemical • Seawater • Composite material • Corrosion process • Scanning electron microscopy (SEM)

---

N. Alvarez-Lemus • R. Orozco-Cruz • R. Galvan-Martinez (✉)
Unidad Anticorrosión, Instituto de Ingeniería, Universidad Veracruzana, Veracruz, México
e-mail: rigalvan@uv.mx

C.A. Leon
Instituto de Investigaciones Metalúrgicas, Universidad Michoacana, Morelia, Mich., México

A. Contreras
Instituto Mexicano del Petróleo, Eje Central Lázaro Cárdenas Norte #152, Col. San Bartolo Atepehuacan, Gustavo A. Madero 07730, México

 
R. Pérez Campos et al. (eds.), *Materials Characterization*,
DOI 10.1007/978-3-319-15204-2_15

## 15.1 Introduction

The manufacture of new materials with superior features to the traditional metals and alloys has been the main trigger for the development of composites materials with metallic matrix or also called composites or MMCs (metal matrix composites). These MMCs have been displacing to the commercial steels, mainly when specific properties are needed [1]. The greatest difficulty in the composite manufacture at liquid state is the wettability of the ceramic phase on the liquid metals [2]. According to this fact, it is important to improve the wettability in order to get a good junction between the metallic matrix and the ceramic [3]. To improve the wettability the addition of surface-active elements to the metal is used; suitable alloying elements in the melt may promote wetting by reducing the surface tension, decreasing the solid/liquid interfacial tension or reacting at the liquid/solid interface [4, 5]. The effect of alloying elements on the wetting behavior of TiC substrates by commercial aluminum alloys (1010, 2024, 6061, 7075) and its relation to phase formation at the metal–ceramic interface was investigated by Aguilar et al. [6]. Some other studies of TiC with metallic matrix of Al–Cu alloy with different content of copper were carried out elsewhere [7]. The high values of activation energies indicated that spreading is not a simple viscosity controlled phenomenon but is a chemical reaction process.

Aluminum alloys are the most common matrix employed in MMCs, it is reported that the addition of TiC as a reinforcement improves the mechanical properties and the resistance at high temperature [8–10]. The composites of the Al/TiC system have been easily fabricated by capillary melt infiltration [11].

Albiter et al. [12] studied the corrosion resistance of composites in 3.5 % NaCl solution evaluated as a function of the addition of Cu and Mg into the aluminum. In all cases pitting corrosion was observed with or without additions of Cu or Mg, but these elements increased the anodic corrosion current. In a similar way, the corrosion behavior of composites reinforced with different ceramics, such as $Al_2O_3$, SiC, and TiC has been studied by Deuis et al. [13] in 3.5 wt.% sodium chloride solution. They found that the corrosion rate increases in the following order: $Al_2O_3 < SiC < TiC$. The corrosion rates of composites were higher than their matrix alloys when they were immersed in NaCl solutions.

Seawater is a high corrosive environment and the corrosion of the metallic structure in this electrolyte is influenced by many factors, such as temperature, pH, salinity, etc. [14].

Al–Cu/TiC composite was obtained to satisfy the mechanical properties and, in order to generate information about the corrosion kinetic of this composite, synthetic seawater was used as corrosive environment. It is important to point out that the electrochemical technique potentiodynamic polarization curves and the corrosion potential were used in the corrosion test. A scanning electronic microscope (SEM) was used to carry out the superficial analysis in order to analyze the morphology of the corrosion process.

## 15.2 Experimental

### *15.2.1 Fabrication of Al–Cu/TiC Composites*

Porous preforms were prepared by uniaxially pressing TiC powders (H.C. Starck grade c.a.s.) with an average size particle of $D_{50}=4.2$ µm, in a rectangular die to form green bars of 6.5×1×1 cm in size. These preforms were sintered at 1,250 °C for 1 h in a tube furnace under flowing argon, obtaining porous preforms with 60 % of theoretical density. The aluminum alloy (Al–4 % Cu) was used to infiltrate the porous preforms at 1,100 °C in an argon atmosphere. A thermogravimetric technique was used to analyze the infiltration process. The final Al–Cu/TiC composites have a volumetric fraction of about 60 % of TiC and 40 % of Al–Cu alloy.

### *15.2.2 Samples Preparation*

Composite samples (working electrode, WE) used in all tests were machined from the Al–Cu/TiC composite (called "composite") and Al–Cu alloys (called "alloy"), the total exposed area of this WE was 1 $cm^2$. Before each test, the WE was polished with silicon carbide (SiC) paper up to 600 grit, and then it was cleaned with deionized water, degreased with acetone, and kept in desiccators.

### *15.2.3 Test Solution*

Synthetic seawater was used as test environment. This solution was made according to ASTM D1141 [15]. The exposure time in all corrosion tests was 24 h at room temperature and atmospheric pressure (Veracruz Port, México) and static conditions.

### *15.2.4 Experimental Setup*

A three-electrode electrochemical glass cell was used; working, reference, and auxiliary electrode. A saturated calomel electrode (SCE) and a sintered graphite rod were used as reference (RE) and auxiliary electrode (AE), respectively. In order to minimize the effect of the solution resistance a Lugging capillary was used.

### 15.2.5 Electrochemical Measurements

The electrochemical techniques used in the present work in order to get the electrochemical measurements were: corrosion potential ($E_{corr}$) and potentiodynamic polarization curves (PCs).

a) $E_{corr}$ versus time.
b) PCs were recorded at a sweep rate of 0.001 V/s and the potential range used was from +0.5 to –0.5 V referred to $E_{corr}$.

These electrochemical techniques were carried out at several time intervals during 24 h and all electrochemical tests were made on clean samples and in freshly prepared test solutions. In order to get a better answer of the corrosion morphology, the total exposure time was 72 h. After exposure time, selected samples were used to make a superficial analysis through an SEM.

## 15.3 Results and Discussion

### 15.3.1 $E_{corr}$ Versus Time

Figure 15.1 shows the results obtained in the corrosion potential ($E_{corr}$) measurements versus time of the composite and alloy immersed in synthetic seawater at room temperature, atmospheric pressure, and static conditions. The results shown in Fig. 15.1

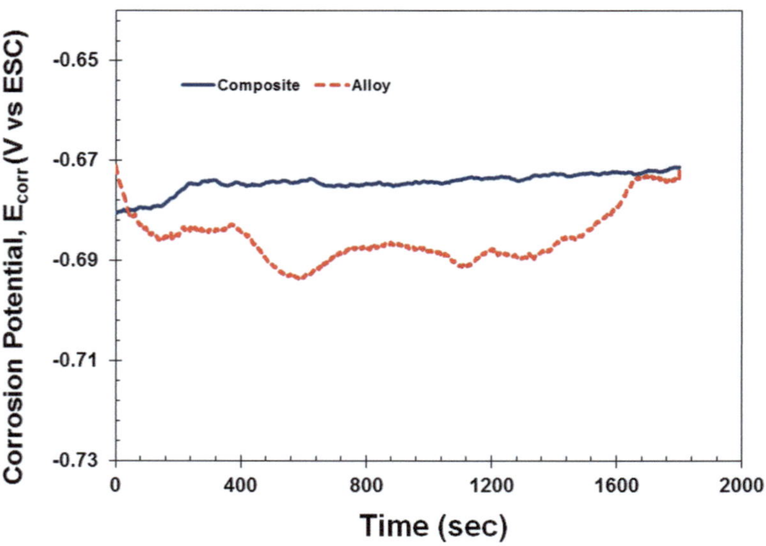


**Fig. 15.1** $E_{corr}$ vs time for Al–Cu/TiC composite and Al–Cu alloy immersed in seawater

indicate the variation of the measured $E_{corr}$ of the composite and alloys immersed in seawater. In this figure it is possible to observe that the $E_{corr}$ values, for the alloy, are slightly more electronegative (–0.69 V approximately) than $E_{corr}$ values corresponding to composite (–0.675 V approximately).

The difference of the $E_{corr}$ values between composite and alloy is really small, it means that the $E_{corr}$ values are similar in both materials. This behavior should be attributed to the fact that the galvanic current presented in the alloy is similar in the composite, because the electrochemical response of this composite is provided only by the metallic matrix.

### *15.3.2 Potentiodynamic Polarization Curves*

Figures 15.2 and 15.3 show the polarization curves of the composite and Al–Cu alloy immersed in seawater, respectively, at room temperature, atmospheric pressure, and static conditions. Figure 15.2 shows the PCs of the composite in seawater. Figure 15.2 shows that at the beginning of the test (0 h), and in the anodic branch, a transfer charge resistance process is limited to the anodic reaction, but at the end of the curve, the anodic Tafel slope increased considerably indicating that a passive corrosion products film is adsorbed on the surface of the composite, whereas in the cathodic branch, a cathodic limiting current density ($i_{LIM}$) is possible to observe. This $i_{LIM}$ is attributed to the oxygen diffusion process, mainly to the diffusion of the oxygen through the corrosion products film [16].

At 12 and 24 h of the exposure time, the polarization curves corresponding to anodic branches have slopes that cannot be associated with a pure charge transfer.

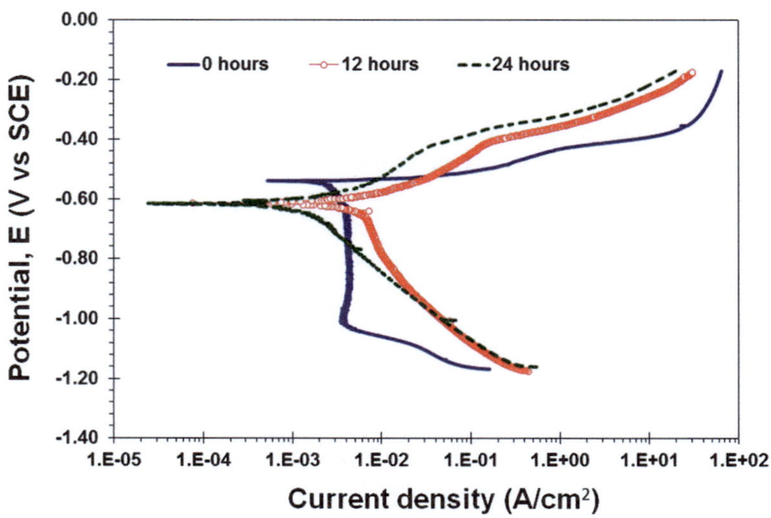


**Fig. 15.2** Polarization curves as a function of the time of the composite immersed in seawater

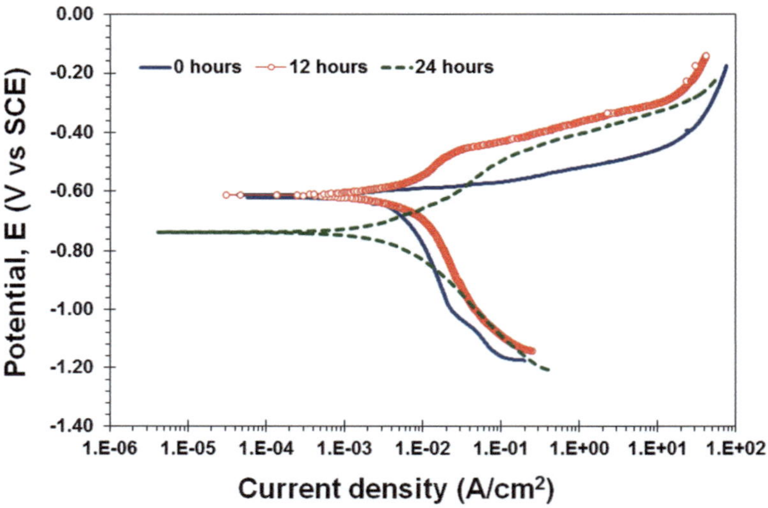


**Fig. 15.3** Polarization curves as a function of the time of the Al–Cu alloy immersed in seawater

**Table 15.1** Electrochemical parameter obtained from the PCs of Fig. 15.2

| Time (h) | ba (V) | bc (V) | $i_{corr}$ (A/cm$^2$) | CR (mm/year) |
|---|---|---|---|---|
| 0 | 0.073 | 1.694 | $9.31 \times 10^{-6}$ | 0.101 |
| 12 | 0.138 | 0.353 | $6.30 \times 10^{-6}$ | 0.068 |
| 24 | 0.168 | 0.270 | $2.44 \times 10^{-6}$ | 0.026 |

This feature suggests a contribution of a mass transfer process on the anodic reaction. On the other hand, in the cathodic branches, the reaction remains limited by the oxygen diffusion process.

Table 15.1 presents the relevant electrochemical parameter calculated from these polarization curves. The parameters are cathodic Tafel slope (bc), anodic Tafel slope (ba), corrosion current density ($i_{corr}$), and corrosion rate (CR). For the present work, the values of the Tafel slope presented in this table were used in all calculations of corrosion rate (CR).

In Table 15.1 it is possible to observe that the highest CR value was obtained at the beginning of the test (0 h). This behavior should be attributed to the fact that at the beginning of the test, the surface of the composite was active, because the surface preparation before this test can eliminate all corrosion products. It is important to point out that the CR decreased as the exposure time increased; it is attributed to the formation of a typical passive corrosion products film of the aluminum under a corrosion process. Figure 15.3 shows the CPs of the polarization curves of the composite and Al–Cu alloy immersed in seawater.

Figure 15.3 shows that at the beginning of the test (0 h), and in the anodic branch, a transfer charge resistance process is limited to the anodic reaction, but at the end of the curve, the anodic Tafel slope increased considerably indicating that a passive corrosion products film is adsorbed on the surface of the alloy. On the other hand, in

**Table 15.2** Electrochemical parameter obtained from the PCs of Fig. 15.3

| Time (h) | ba (V) | bc (V) | $i_{corr}$ (A/cm$^2$) | CR (mm/year) |
|---|---|---|---|---|
| 0 | 0.069 | 0.405 | $1.46 \times 10^{-6}$ | 0.016 |
| 12 | 0.156 | 0.448 | $3.37 \times 10^{-6}$ | 0.037 |
| 24 | 0.131 | 0.251 | $2.03 \times 10^{-6}$ | 0.022 |

the cathodic branch, a not well-defined cathodic limiting current density ($i_{LIM}$) can be observed. The behavior in the anodic and cathodic reactions is similar to the behavior presented in the alloy. The $i_{LIM}$ is attributed to the oxygen diffusion process.

At 12 and 24 h of the exposure time, the results of the alloy are similar to the composite (Fig. 15.2), the polarization curves corresponding to anodic branches have slopes that cannot be associated with a pure charge transfer. This feature suggests a contribution of a mass transfer process on the anodic reaction. On the other hand, in the cathodic branches, the reaction remains limited by the oxygen diffusion process [17].

The results shown in Figs. 15.2 and 15.3 indicate that the corrosion processes of the alloy and composite are provided by the galvanic couple [18] Al–Cu and specifically by the Al.

Table 15.2 presents the relevant electrochemical parameter calculated from the polarization curves of Fig. 15.3. For the present work, the values of the Tafel slope presented in this table were used in all calculations of CR.

Table 15.2 shows that the highest CR value was obtained at the middle of the test (12 h). This behavior should be attributed to the fact that the passive corrosion products film formed on surface of the alloy was broken (highest CR values) by the chloride ions actions, but the CR decreased at 24 h indicating that the passive film was regenerated. It is clear the influence of TiC on the Al–Cu alloy on the composite, because the lowest CR values correspond to the alloy sample.

Figure 15.4 shows the PCs of the composite and Al–Cu alloy at the beginning of the test and after 24 h of the exposure time in seawater.

Figure 15.4 shows graphically that the corrosion current density is lower in the two PCs of the Al–Cu alloy than in the two CPs of the composite. In all PCs it is possible to observe the influence of a diffusion process attributed mainly to the corrosion products film formed on surface of the material sample. It is important to point out that aluminum is the matrix in the alloy and composite, for that reason the corrosion products (aluminum oxide) should be a passive film formed on surface of the total exposure area. These results are correlated with the corrosion morphology that is shown in Fig. 15.5. According the pitting type mechanism, it is necessary the broken of the passive film and then get an aeration differential cell. In addition, a mix process is possible to observe in all PCs.

### *15.3.3 Surface Analysis by SEM*

Figure 15.5 shows the micrographs obtained by SEM from the selected composite and Al–Cu alloy samples exposed to seawater after 72 h of the exposure time.

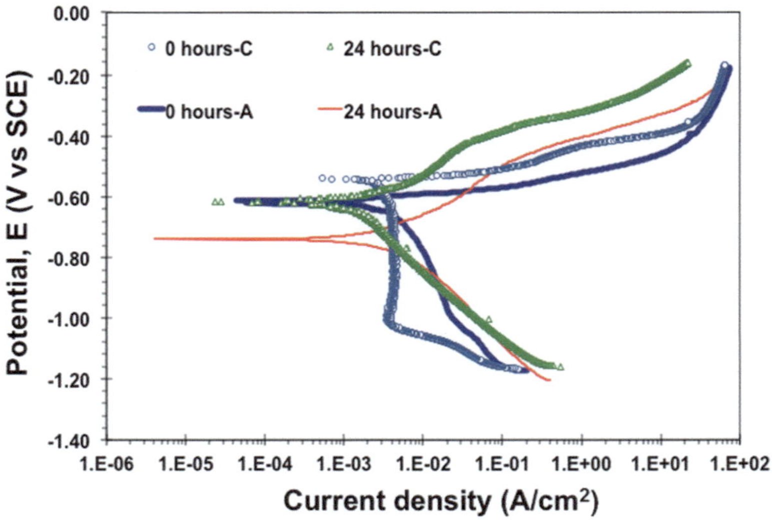


**Fig. 15.4** Polarization curves as a function of the time of the composite (0 and 24 h-C) alloy (0 and 24 h-A) immersed in seawater

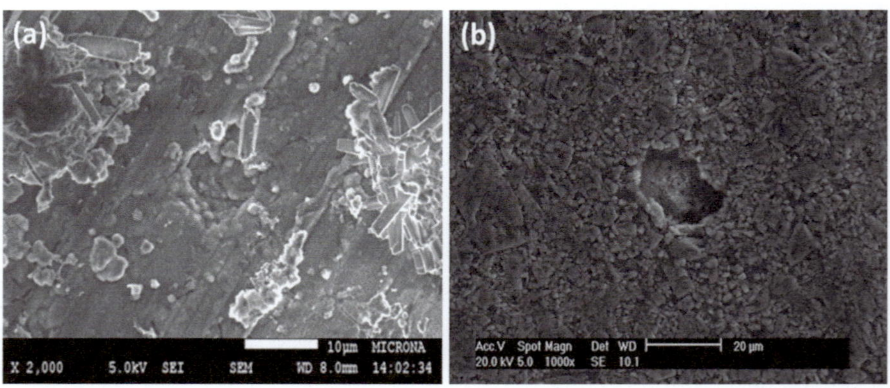


**Fig. 15.5** SEM micrographs of the morphological attack by corrosion of samples immersed in seawater during 72 h (**a**) Al–Cu alloy (**b**) Al–Cu/TiC composite

In the two micrographs corresponding to Al–Cu alloy (a) and composite (b) it is possible to observe that the corrosion form is localized, specifically, pitting corrosion. It is important to point out that the mechanism that limited this pitting corrosion type is the aeration differential and it was caused by the rupture of corrosion products film (the passive film) adsorbed on surface of the metallic sample. These results are agreed with the corrosion form found by Gavgali et al. [19].

## 15.4 Conclusions

According to electrochemical study of the composite and alloy sample exposed to seawater at static conditions, room temperature, and atmospheric pressure, it is possible to conclude that the corrosion rate values of the Al–Cu/TiC composite were higher than the CR values of the Al–Cu alloy; this fact is attributed to the titanium carbide that it should induce to the crevice corrosion form. The total corrosion of the composite and alloy is a mix process, because a charge transfer resistance corresponding to a pure corrosion process is limited by a diffusional resistance (diffusion polarization induced by the oxygen diffusion process through the corrosion products film). The mechanism of the alloy and composite corrosion is a differential aeration cell, where it can produce pitting or crevice corrosion form. The corrosion form found for the composite and Al–Cu alloy samples in the test solution was a localized corrosion, specifically pitting corrosion.

**Acknowledgments** The authors would like to thank the National Council of Science and Technology (CONACyT) for the grant awarded to Miss N. Alvarez-Lemus, required to develop this work.

## References

1. Candan S (2009) An investigation on corrosion behaviour of pressure infiltrated Al–Mg alloy/SiCp composites. Corros Sci 51:1392–1398
2. Contreras A, Albiter A, Bedolla E, Perez R (2004) Processing and characterization of Al–Cu and Al–Mg based composites reinforced with TiC. Adv Eng Mater 6(9):767–775
3. Contreras A, Leon CA, Drew RAL, Bedolla E (2003) Wettability and spreading kinetics of Al and Mg on TiC. Scr Mater 48:1625–1630
4. Delannay F, Froyen L, Deruyttere A (1987) The wetting of solids by molten metals and its relation to the preparation of metal–matrix composites. J Mater Sci 22:1–16
5. Banerji A, Rohatgi PK, Reif W (1984) Role of wettability in the preparation of metal–matrix composites (a review). Metall Tech 38:656–661
6. Aguilar EA, Leon CA, Contreras A, Lopez VH, Drew RAL, Bedolla E (2002) Wettability and phase formation in TiC/Al-alloys assemblies. Compos Part A 33:1425–1428
7. Contreras A (2007) Wetting of TiC by Al–Cu alloys and interfacial characterization. J Colloid Interface Sci 311:159–170
8. Kennedy AR, Wyatt SM (2000) The effect of processing on the mechanical properties and interfacial strength of aluminum/TiC MMCs. Compos Sci Technol 60:307–314
9. Muscat D, Shanker K, Drew RAL (1992) Al/TiC composites produced by melt infiltration. Mater Sci Technol 8:971–976
10. Albiter A, Leon CA, Drew RAL, Bedolla E (2000) Microstructure and heat treatment response of Al-2024/TiC composites. Mater Sci Technol 289:109–115
11. Contreras A, Salazar M, León CA, Drew RAL, Bedolla E (2000) Kinetic study of the infiltration of aluminum alloys into TiC preforms. Mater Manuf Processes 15:163–182
12. Albiter A, Contreras A, Salazar M, Rodriguez JG (2006) Corrosion behavior of aluminum metal matrix composites reinforced with TiC processed by pressureless melt infiltration. J Appl Electrochem 36:303–308

13. Deuis RL, Green L, Subramanian C, Yellup JM (1997) Corrosion behavior of aluminum composite coatings. Corrosion 53:880–890
14. Silverman DC (2003) Aqueous corrosion. In: Cramer SD, Covino BS (Eds) Corrosion: Fundamentals, Testing and Protection, Vol 13A. ASM Handbook, ASM International, Materials Park, OH, pp190–195
15. ASTM D1141 (2013) Standard practice for the preparation of substitute ocean water
16. Galvan-Martinez R, Cabrera D, Galicia G, Orozco R, Contreras A (2013) Electrochemical characterization of the structural metals immersed in natural seawater: "in situ" measures. Mater Sci Forum 755:119–124
17. Galvan-Martinez R, Mendoza-Flores J, Duran-Romero R, Genesca J (2007) Effect of turbulent flow on the anodic and cathodic kinetics of API X52 steel corrosion in $H_2S$ containing solutions: a rotating cylinder electrode study. Mater Corros 58:514–521
18. Budruk AS, Balasubramaniam R, Gupta M (2008) Corrosion behavior of Mg–Cu and Mg–Mo composites in 3.5 % NaCl. Corros Sci 50:2423–2428
19. Gavgali M, Dikici B, Bedir F (2007) Corrosion susceptibilities of Al–Cu MMCs fabricated by conventional hot pressing. Indian J Eng Mate Sci. 14:303–308

# Part VI
# Characterization of Materials for Medical Applications

This part includes two chapters related with studies of materials for medical applications.

In the first chapter, $Fe_3O_4$ magnetic nanoparticles, which are used for the diagnostic and treatment of diseases, were synthesized via co-precipitation of ferrous and ferric ions in alkali solution. The morphology and the structure of the nanoparticles were characterized using aberration-corrected scanning transmission electron microscopy. The contrast of the high angle annular dark field images indicated the contribution mainly of the Fe atoms given that the contribution of the O atoms is very weak. This was checked by simulated images.

In the second chapter, the synthesis and physicochemical characterization of Mg/Al hydrotalcite as well as its arsenic removal capacity is described. The adsorbent materials were synthesized by the co-precipitation method and characterized to identify its structure. The results shown that Mg/Al hydrotalcite phases have a great capability to remove arsenic from aqueous systems in comparison with Fe-oxides-hydroxides, which are generally considered between the most efficient arsenic removers.

The main techniques used in the characterization of these materials are:

- Aberration-corrected scanning transmission electron microscopy (STEM).
- X-ray diffraction (XRD).
- Scanning electron microscopy (SEM).
- Fourier transform infrared (FTIR).
- Differential thermal analysis (DTA).
- Thermo gravimetric analysis (TGA).

# Chapter 16
# Adsorption of Arsenite from Aqueous Solution on Mg/Al Hydrotalcite

**E. Ramos-Ramírez, Norma Leticia Gutiérrez-Ortega, R. Zarraga-Núñez, and F.J. Acevedo-Aguilar**

**Abstract** In the present study the synthesis and physicochemical characterization of Mg/Al hydrotalcite as well as its arsenic removal capacity is described. The adsorbent materials were synthesized by the coprecipitation method and characterized to identify its structure. Adsorption isotherm studies of As (III) from aqueous solution are described using the Langmuir, Freundlich, and Dubinin–Radushkevich models. The As (III) adsorption isotherm data fit best to the Freundlich model. The maximum As (III) adsorption efficienciesare 791.92 mg As (III)/g Mg/Al Hydrotalcite = 2, 642.56 mg As (III)/g Mg/Al Hydrotalcite = 3, 963.79 mg As (III)/g Mg/Al Hydrotalcite = 2 at 350 °C and 827.54 mg As (III)/g Mg/Al Hydrotalcite = 3 at 350 °C, in contact times less than 2 min. These results shown that Mg/Al hydrotalcite phases have a great capability to remove arsenic from aqueous systems in comparison with Fe-oxides-hydroxides, which are generally considered between the most efficient arsenic removers. This capacity can be attributed to a combination of anionic exchange, surface capture, and precipitation processes occurring over Mg/Al hydrotalcite, which made it as a potential material to remove arsenite efficiently.

**Keywords** Hydrotalcite • Adsorption • Arsenite • Aqueous solution • Removers

## 16.1 Introduction

Arsenic contamination of natural water is a worldwide crisis from the last decades. Arsenic is a semi-metallic element, which rarely is found free in nature. Generally, arsenic is found in the environment as complex mineralogical forms. Arsenic as

E. Ramos-Ramírez (✉) • R. Zarraga-Núñez • F.J. Acevedo-Aguilar
Divisi6n de Ciencias Naturales y Exactas, Departamento de Química, de la Universidad de Guanajuato, Noria Alta s/n, Col. Noria Alta, Guanajuato, GTO 36050, México
e-mail: ramosre@ugto.mx

N.L. Gutiérrez-Ortega
Divisi6n de Ingenierías, Departamento de Ingeniería Civil, de la Universidad de Guanajuato, Juárez No. 77, Col. Centro, Guanajuato, GTO 36000, México
e-mail: normagut@ugto.mx

 
R. Pérez Campos et al. (eds.), *Materials Characterization*,
DOI 10.1007/978-3-319-15204-2_16

trivalent and pentavalent forms are extensively found in natural aqueous systems, but As (III) is about 60 times more toxic and has high mobility than As (V). Contamination of drinking water with arsenic occurs from mineral weathering, volcanic emissions, and anthropogenic process. Arsenic compounds were widely used since the 1970s as pesticides, fertilizers, and wood preservers, also as component of glass, non-ferrous alloys, solar cells, light-emitting diodes, lasers, integrated circuits, and also used in medicine [1]. Epidemiological studies and case reports of humans exposed to arsenic in drinking water have demonstrated that acute exposition effects include fever, vomiting, severe diarrhea, dehydration, hepatomegaly, cardiac arrhythmia, upper respiratory tract symptoms, and peripheral neuropathies. Ingestion of large doses of inorganic arsenic such as 70–180 mg may be lethal. In addition, its exposure increases the risk of cancer in the skin, lung, digestive tract, liver, bladder, kidney, and lymphatic and hematopoietic systems. Therefore, arsenic was reviewed for listing in the First Annual Report on Carcinogens and by the International Agency for Research on Cancer [2].

Several methods are used nowadays to remedy arsenic contamination. Those treatment methods are precipitation, solvent extraction, nanofiltration, reverse osmosis ion-exchange, membrane process, electrochemical reduction, and adsorption. Each of the above processes has its own advantages and disadvantages, which make the difficulty to select a suitable process. The disadvantages of traditional methods include a very limited ability to remove arsenic because of exchange competition from other anions found in groundwater (ion-exchange), high sludge production (coagulation–flocculation), membrane fouling (nanofiltration), constant monitoring of the ions concentration (ion-exchange) and high cost (membrane processes, electrochemical reduction, and reverse osmosis). Comparing all the disadvantages of abovementioned processes, adsorption processes are effective techniques used in the water and wastewater industries being explored for its easy handling, minimal sludge production, and regeneration capability. Different natural and synthetic materials are used to remove arsenic. Some of these are biosorberts (agriculture wastes or biological tissues), natural minerals (kaolinite, magnetite,maghemite, hematite, and feldspar) or synthetic material (activated coal, titania, zeolites, alumina, iron oxides, titanium oxides and hybrid polymeric.) [3–5].

Another group of synthetic adsorbents are hydrotalcites, which present a great potential for removal of contaminants and present low cost. Hydrotalcites are a class of layered compound derived from mineral brucite, in which magnesium is octahedrally surrounded by hydroxyl ions and different octahedrals share edges to form infinite sheets. When some divalent cations are substituted by trivalent cations, a positive charge is created on the layers, which is compensated by anions and water molecules in the interlayer zone, resulting in an excellent capacity to capture anionic contaminants [6].

In previous works developed by our group hydrotalcites obtained by the sol–gel process have been used to remove chromate and arsenite ions from aqueous systems showing a great capacity [7, 8], but presenting the disadvantage of being very costly. For which in this work this presents the results obtained in the processes of adsorption of arsenite using hydrotalcites obtained by the coprecipitation method.

## 16.2 Experimental

### *16.2.1 Preparation of Mg/Al Hydrotalcites*

The hydrotalcite with Mg/Al = 2 ratio was synthesized using the coprecipitation method: 0.5 mol of $Mg(NO_3)_2{\cdot}6H_2O$ and 0.25 mol of $Al(NO_3)_3{\cdot}9H_2O$ were dissolved in water. The solution was coprecipitated with NaOH under constant stirring. The pH of the mixture was maintained at 11.5. The sample was aged for 24 h. After that, it was rinsed with water until reaching a pH of 8.0 and finally was dried at 70 °C for 24 h. To obtain hydrotalcite of Mg/Al = 3 ratio were used 0.75 mol of $Mg(NO_3)_2{\cdot}6H_2O$ and 0.25 mol of $Al(NO_3)_3{\cdot}9H_2O$ in the same synthesis conditions. The solids synthesized HTCP-2 and HTCP-3 also were heated treated at 350 °C to obtain the HTCP-2 350 °C and HTCP-3 350 °C, respectively.

### *16.2.2 Characterization of Mg/Al Hydrotalcites*

The solids were characterized to determine its thermal, structural and textural properties. Scientific instruments utilized: SDT Q600 Instruments at heating rate of 10 °C/min in air atmosphere at 100 mL/min; INEL Equinox Diffractometer (powders) coupled to a copper anode tube using monocromatized CuK$\alpha$ radiation; Automatized Micromeritics ASAP 2010; Fourier Transform Infrared (FTIR) Perkin-Elmer 1600 series in a range of wave length of 4,000–400 $cm^{-1}$; and JEOL JSM 5900LV low vacuum Scanning Electron Microscopy.

### *16.2.3 Adsorption Isotherms*

Adsorption isotherms were determined in a batch procedure at 25 °C and stirred by a rotator by addition of 0.10 g of adsorbent into 10 mL of As (III) solutions with arsenic concentrations ranging from 50 to 400 mg/L. The equilibrium As (III) concentrations were determined using an atomic absorption spectroscopy (Perkin Elmer Analyst 200 with an Arsenic hollow cathode lamp at $\lambda = 193.7$ nm). The experimental results obtained of the adsorption isotherms were analyzed by the models of Langmuir, Freundlich and Dubinin–Radushkevich. Adsorption parameters and correlation coefficients of the curves were calculated from the adsorption isotherm data.

## 16.3 Results and Discussion

### *16.3.1 Characterization of Mg/Al Hydrotalcites*

The XRD patterns for HTCP-2 and HTCP-3 non-calcined (Fig. 16.1) were characteristics for a mineralogical phase type hydrotalcite (JCPDS 22-700), with sharp and intense peaks at lower $2\theta$ values and less intense peaks at higher $2\theta$. Slight shifts were

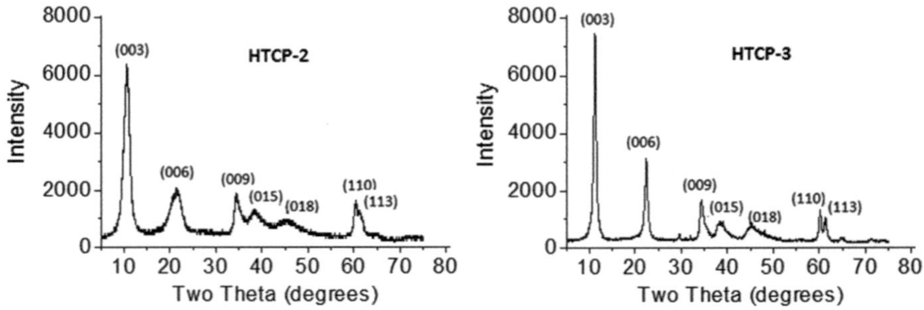


**Fig. 16.1** X-ray diffraction patterns for non-calcined Mg/Al hydrotalcites

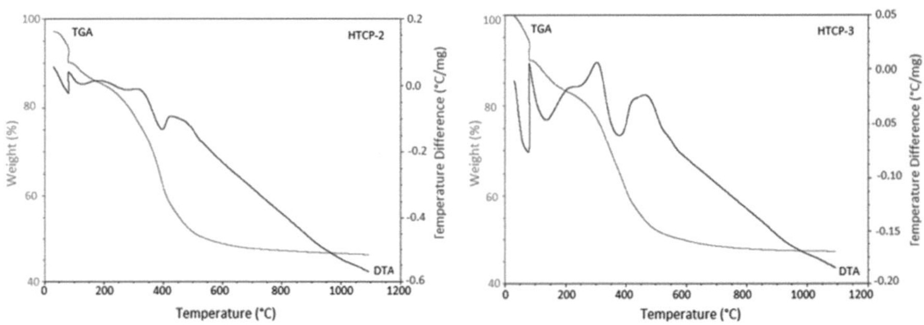


**Fig. 16.2** DTA (temperature difference *black lines*) and TGA (weight *gray lines*) curves for Mg/Al hydrotalcites

observed in major peaks towards higher 2$\theta$ with Mg/Al=3 molar ratio which also indicates higher charge density and also a decreased interlayer distance and unit cell parameters. This difference in interlayer distance could be due to a decrease in the electrostatic interaction between positive layers and negative interlayers [9].

Thermal decomposition for both hydrotalcites (DTA and TGA) is shown in Fig. 16.2. HTCP-2, in the temperature range 30–250 °C evidenced endothermic effects attributed to physisorbed and interlayered water loss. Between 250–450 °C occur decarbonation and partial dehydroxylation to produce $Mg_6Al_2O_8(OH)_2$ phase. At temperatures below 450 °C mixed phases of periclase (MgO) and spinel ($MgAl_2O_4$) are found with loss weight of 34 % [10].

Thermal decomposition patron is shown in the following reactions sequence:

$$Mg_6Al_2\left(OH\right)_{16}CO_3\cdot 4H_2O \rightarrow Mg_6Al_2\left(OH\right)_{16}CO_3 + 4H_2O\uparrow \; 25 \quad 250\ C \qquad (16.1)$$

$$Mg_6Al_2\left(OH\right)_{16}CO_3 \rightarrow Mg_6Al_2O_8\left(OH\right)_2 + 7H_2O\uparrow + CO_2\uparrow \; 250-450°C \qquad (16.2)$$

$$Mg_6Al_2O_8\left(OH\right)_2 \rightarrow MgAl_2O_4 + 5MgO + H_2O\uparrow \qquad 450-900°C \qquad (16.3)$$

In Fig. 16.3 are shown the XRD patterns of HTCP-2 and HTCP-3 heated at 350 °C, showing the presence of periclase (Card JCPDS-4-0829) and hydrotalcite phases.

The FTIR spectra for the four solids are shown in Fig. 16.4, which are typical for hydrotalcites. The main difference in the spectra is the band at 1,360 cm$^{-1}$ attributed to $\nu_3$ for the $CO_2^{3-}$ anions with $D_3h$ symmetry, which is evidence of the formation of hydrotalcite containing carbonate anions interacting with water molecules that show a band at 1,625 cm$^{-1}$ [11].

In Fig. 16.5 are shown BET isotherms as the calcination temperature affects the surface area and the porosity of the solids. A maximized increase in the surface area

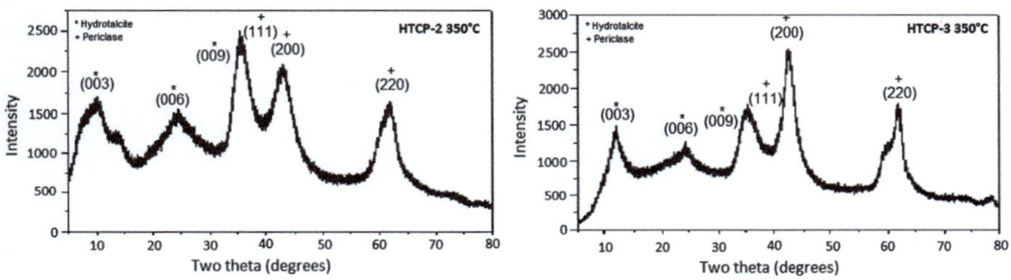


**Fig. 16.3** X-ray diffraction patterns of calcined Mg/Al hydrotalcites at 350 °C

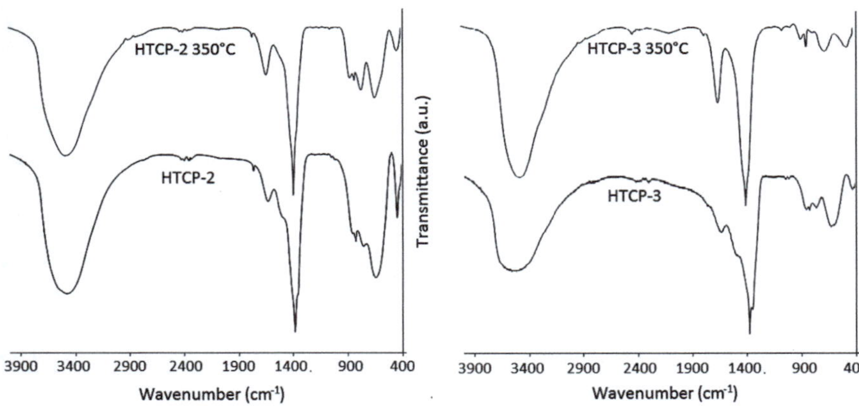


**Fig. 16.4** FTIR spectra of non-calcined and calcined Mg/Al hydrotalcites

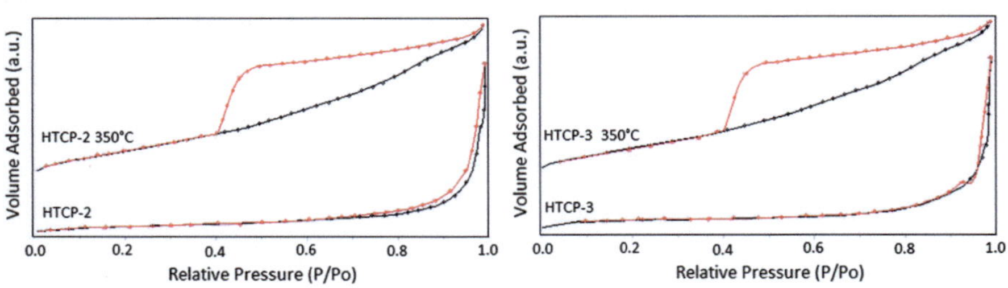


**Fig. 16.5** BET isotherms of uncalcined and calcined Mg/Al hydrotalcites

**Table 16.1** Textural properties of non-calcined and calcined Mg/Al hydrotalcites

| Sample | Surface area ($m^2/g$) | Pore size (nm) | Pore volume ($cm^3/g$) | Isotherm type | Hysteresis type |
|---|---|---|---|---|---|
| HTCP-2 | 9.9 | 18.75 | 0.0471 | V | H3 |
| HTCP-3 | 1.6 | 13.42 | 0.0052 | V | H3 |
| HTCP-2 350 °C | 95.2 | 7.22 | 0.1719 | IV | H4 |
| HTCP-3 350 °C | 75.1 | 6.17 | 0.1158 | IV | H4 |

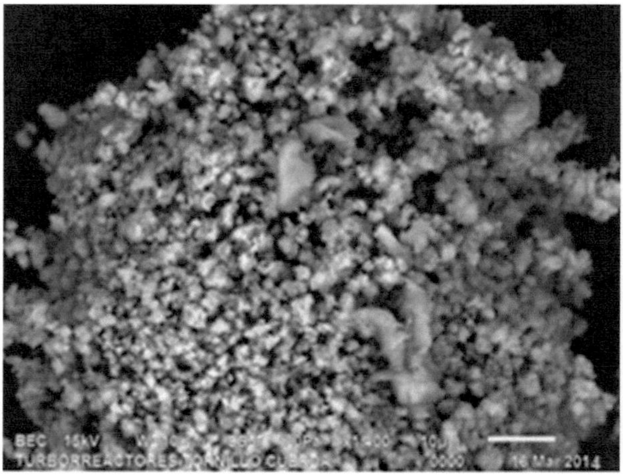

**Fig. 16.6** SEM image uncalcined Mg/Al hydrotalcites

was generally observed for the samples obtained after calcination at 350 °C due to molecules that to scape from structure. The HTCP-2 and HTCP-3 shown an $N_2$ adsorption–desorption type V isotherms/H3 buckle, associated with mesoporous material with low pore volume and slit type porosity. For HTCP-2 and HTCP-3 at 350 °C are observed Type IV isotherms H4 buckle corresponding to mesoporous materials with big pore volume and narrow slit porosity [12]. Porosity increased by the change in crystalline structure due to thermal evolution.

In Table 16.1 are presented textural properties of Mg/Al hydrotalcites. The surface area values is increased with the thermal treatment due to water, carbonates, and hydroxide loss reaching values around of 100 $m^2/g$ associated with mesoporosity (very important property for the adsorption process).

Scanning electron microscope (SEM) image of Mg/Al hydrotalcite shows that well-defined particles with a homogeneous particle size distribution forms aggregate constituted by smaller particles (Fig. 16.6).

### *16.3.2 Adsorption Isotherms*

Initial concentration is one of the main factors to determine the adsorption isotherm. The experimental results of the effect of initial arsenic concentration on removal efficiency are presented in Fig. 16.7. The As (III) removal efficiency increases with

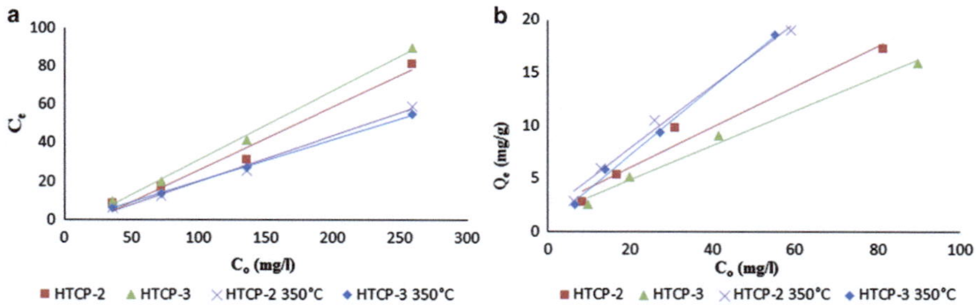


**Fig. 16.7** Effect of initial arsenic concentration on removal efficiency of As (III) into non-calcined and calcined Mg/Al hydrotalcites

the increase in initial concentration. The above-mentioned relation is confirmed providing a directly proportional behavior between the initial concentration of As (III) adsorbed in a certain quantity of Mg/Al hydrotalcites as adsorbent material.

The relation between the amount of As (III) adsorbed by the hydrotalcite and the equilibrium concentration of As (III) can be expressed by the linearized Langmuir adsorption isotherm as:

$$\frac{1}{Q_e} = \frac{1}{Q_{max}} + \frac{1}{bQ_{max}} \frac{1}{C_e} \tag{16.4}$$

where $Q_e$ is the amount of As (III) adsorbed at equilibrium, $C_e$ is the As (III) concentration at equilibrium in the solution, $Q_{max}$ is the maximum adsorption intensity of As (III), and $b$ is the adsorption constant related to the enthalpy of adsorption.

The linearized Freundlich isotherm is expressed as:

$$\ln Q_e = \ln K_f + \frac{1}{n} \ln C_e \tag{16.5}$$

where $K_f$ represent the adsorption capacity; $n$ represents the adsorption intensity.

The linearized Dubinin–Radushkevich estimates the adsorption energy as:

$$\ln Q_e = \ln Q_D - B_D \left[ RT \left( 1 + \frac{1}{C_e} \right) \right]^2 \tag{16.6}$$

where $B_D$ is related to the free energy of adsorption per mole of As (III) as it migrates to the surface of the adsorbent from infinite distance in the solution, and $Q_D$ is the Dubinin–Radushkevich isotherm constant related to the degree of As (III) sorption by the hydrotalcite surface [7].

Table 16.2 showed the values of maximum removal capacity, factor of correlation, and equilibrium constants to Langmuir, Freundlich, and Dubinin–Radushkevich isotherm models obtained by linear plots.

The values of $R^2$ showed that the As (III) adsorption by Mg/Al hydrotalcites does not fit well to the Langmuir adsorption model. The models that present best correlation are Freundlich and Dubinin isotherm. The linearization for hydrotalcites fits

**Table 16.2** Adsorption isotherms values of As (III) removal onto Mg/Al hydrotalcites

| | Langmuir | | | Freundlich | | | Dubinin–Radushkevich | | | |
|---|---|---|---|---|---|---|---|---|---|---|
| Isotherms sample | $R^2$ | $Q_{max}$ (mg/g) | $K$ (L/mg) | $R^2$ | $K_f$ (mg/g) | $n$ | $R^2$ | $Q_D$ (mg/g) | $B_D$ (mol²/ kJ²) | $E_a$ (kJ/mol) |
| HTCP-2 | 0.99 | 41.6 | 0.009 | 0.99 | 791.9 | 1.2 | 0.99 | 35.4 | 0.028 | 4.23 |
| HTCP-3 | 0.99 | 43.5 | 0.006 | 0.99 | 642.6 | 1.1 | 0.99 | 33.5 | 0.027 | 4.23 |
| HTCP-2 350 °C | 0.98 | 55.5 | 0.501 | 0.99 | 936.8 | 1.1 | 0.99 | 36.6 | 0.028 | 4.15 |
| HTCP-3 350 °C | 0.74 | 90.9 | 0.417 | 1.00 | 827.5 | 1.1 | 0.99 | 37.0 | 0.029 | 4.23 |

reasonably according to Freundlich adsorption equation with high linear regression values ($R^2$) and good removal capacity of As (III) for all solids. Best capacity occurs in the solid Mg/Al = 2 due to the higher amount of aluminum in the sheets generates more positive charge to attract anions. As(III) removal capacity of hydrotalcite is improved after thermal treatment for both Mg/Al relations. With respect to surface area dependence, it is not directly related since fresh solids show low surface values but good As(III) removal capacity.

The $Q_D$ values for the Mg/Al hydrotalcites indicate that these solids exhibit high adsorption capacity. The apparent adsorption energy ($E$) was calculated using the equation:

$$E = \frac{1}{\sqrt{2B_D}} \tag{16.7}$$

These values suggest that the adsorption occurs through a physisorption mechanism also called nonspecific adsorption as a result of long range weak Van der Waals forces between arsenite and hydrotalcite with surface coverage on this experiments fitted best to Freundlich isotherm as sorption on a heterogeneous surface present at adsorption is not restricted to monolayer coverage. The removal process resulted from a combination of anionic exchange, surface capture, and precipitation process over Mg/Al hydrotalcite attributed to the mesoporous-crystalline structure and its capacity to capture anions in the interlamellar space.

## 16.4 Conclusions

In this work, we present an efficient methodology for removal of As (III) in aqueous solutions using hydrotalcites synthesized by coprecipitation method. The materials synthesized present good thermal, textural, and structural properties with the capacity to remove relative high amounts of As (III) efficiently in short time. The results shown that these hydrotalcites have high removal capacity for arsenite. This can be attributed to a combination of anionic exchange, surface capture, and precipitation processes over Mg/Al hydrotalcite. The isotherms of adsorption suggest that the adsorption is favored by a physisorption mechanism which promotes a spontaneous and rapid process. Such a removal capacity increases after thermal treated to a suitable temperature. The results presented here suggest that hydrotalcite synthesized by coprecipitation method may provide cost-reducing alternative in treating wastewater, drinking water, or groundwater containing harmful oxyanions of arsenic.

**Acknowledgments** We are grateful for him to the DAIP and PROMEP for economical support, and I. V. Hernández-Cervantes, and I. Castro-Becerra of Guanajuato University and Dr. G. Herrera of ITESI for the technical support offered for the accomplishment of this work.

## References

1. Guha DN (2008) Chronic arsenic toxicity and human health. Indian J Med Res 128(4):436–447
2. Burwell SM (2014) Report on Carcinogens U.S., 13th ed. Department of Health and Human Services Secretary, Atlanta
3. Mohan D, Pittman CU Jr (2007) Review arsenic removal from water/wastewater using adsorbents—a critical review. J Hazard Mater 142:1–53
4. Ali I, Gupta VK, Khan TA, Asim M (2012) Removal of arsenate from aqueous solution by electro-coagulation method using Al-Fe electrodes. Int J Electrochem Sci 7:1898–1907
5. Han DS, Song JK, Batchelor B, Abdel-Wahab A (2012) Removal of arsenite(As(III)) and arsenate(As(V)) by synthetic pyrite ($FeS_2$): synthesis, effect of contact time, and sorption/desorption envelopes. J Colloid Interface Sci 392:311–318
6. Jobbágy M, Regazzoni AE (2012) Complexation at the edges of hydrotalcite: the cases of arsenate and chromate. J Colloid Interface Sci 8:314–318
7. Ramos-Ramírez E, Gutierrez-Ortega NL, Contreras-Soto C, Olguin-Gutiérrez MT (2009) Adsorption isotherm studies of chromium (VI) from aqueous solutions using sol–gel hydrotalcite-like compounds. J Hazard Mater 172:1527–1531
8. Ramos-Ramírez E, Gutiérrez-Ortega NL, Rangel-Porras G, Herrera-Pérez G (2014) Nano-Mg/Al hydrotalcite: Physicochemical characterization and removal of As(III) from aqueous solutions. Mater Res SocSympProc 1616
9. Cavani F, Trifiro F, Vaccari A (1991) Hydrotalcite-type anionic clays: preparation, properties and applications. Catal Today 11:173–301
10. Wang Q, Huang-Tay H, Guo Z, Chen L, Liu Y, Chang J, Zhong Z, Luo J, Borgna A (2012) Morphology and composition controllable synthesis of Mg–Al–$CO_3$ hydrotalcites by tuning the synthesis pH and the $CO_2$ capture capacity. Appl Clay Sci 55:18–26
11. Roelofs JCAA, Bokhoven JA, Dillen AJ, Geus JW, Jong KP (2002) The thermal decomposition of Mg-Al hydrotalcites: effects of interlayer anions and characteristics of the final structure. Chem Eur J 8(24):5571–5579
12. Rouquero F, Rouquero J, Sing K (1999) Adsorption by powders and porous solids: principles methodology and applications. Academic, San Diego

# Chapter 17
# Synthesis and Characterization of Magnetic Nanoparticles for Biomedical Applications

**J.R. Piñón-Hernández, I.G. Becerril-Juárez, A. Ángeles-Pascual, R. Pérez, and R. Esparza**

**Abstract** Currently, magnetic nanoparticles show a great variety of applications in the biomedicine area, where they are used for the diagnostic and treatment of diseases by means of transportation and release of drugs, immobilization and magnetic separation of biological entities or treatment of tumors through hyperthermia therapies. An inevitable problem associated with magnetic nanoparticles is their intrinsically instability through time. Besides, the particles without covering are highly reactive chemically and they oxidize quickly in the air, producing a loss of magnetism and their ability of dispersion. For this reason it is crucial to develop strategies of protection during or after the synthesis, to chemically stabilize the magnetic particles against degradation. Magnetic nanoparticles ($Fe_3O_4$) were synthesized via co-precipitation of ferrous ($Fe^{2+}$) and ferric ($Fe^{3+}$) ions in alkali solution. The morphology and the structure of the nanoparticles were characterized using aberration-corrected (Cs) scanning transmission electron microscopy (STEM). Comparisons between the theoretical and experimental STEM images were carried out and some insights from the atomic structure of the magnetic nanoparticles are shown. Magnetic nanoparticles were obtained with a homogeneous dispersion and average particle size of 19.5 nm.

**Keywords** Nanoparticles • Synthesis • Biomedical • Coprecipitation • Aberration corrected

J.R. Piñón-Hernández • R. Pérez • R. Esparza (✉)
Centro de Física Aplicada y Tecnología Avanzada, Universidad Nacional Autónoma de México, Boulevard Juriquilla 3001, Santiago de Querétaro, QRO 76230, México
e-mail: resparza@fata.unam.mx

I.G. Becerril-Juárez
Program on Nanoscience and Nanotechnology, CINVESTAV-IPN, Zacatenco, México D.F. 07360, México

A. Ángeles-Pascual
Departamento de Ingeniería Eléctrica-SEES, CINVESTAV-IPN, Zacatenco, México D.F. 07360, México

Programa de Doctorado en Ingeniería y Ciencias Aplicadas, Centro de Investigaciones en Ingeniería y Ciencias Aplicadas, UAEM, Av. Universidad 1001 Col. Chamilpa, Cuernavaca, Morelos, México



R. Pérez Campos et al. (eds.), *Materials Characterization*,
DOI 10.1007/978-3-319-15204-2_17

## 17.1 Introduction

Nanostructured materials are of great interest because of unusual physical and chemical properties, which are due to their size, morphology, and composition. Magnetic nanoparticles such as magnetite ($Fe_3O_4$) have attracted much interest not only in the field of magnetic recording media [1] due to unique electric and magnetic properties based on the transfer of electrons between $Fe^{2+}$ and $Fe^{3+}$ in the octahedral sites, but also in the biomedical applications such as drug delivery systems, magnetic resonance imaging, medical diagnostics, and cancer therapy [2]; in addition, such magnetic nanoparticles have good biocompatibility and low toxicity [3, 4].

However, $Fe_3O_4$ nanoparticles exhibit a tendency to form aggregates due to their magnetic field, which reduces their efficiency considerably, since the nanoparticles surface tends to oxidize. Metallic nanocrystals are easily oxidized with exposure to oxygen or moisture and transformed to metal oxides or hydroxides with inferior magnetic properties [5], compromising their enhancement ability. To prevent the metallic nanocrystals from oxidization and corrosion, compact and chemically stable shells are grown on the nanocrystals. Some progress on the synthesis of magnetic nanoparticles has been focused on the stabilized nanoparticles to prevent agglomeration, oxidation and induce chemical functionality, mainly using the surfactants or capping ligands with some specific functional groups [6]. And also, many noble metals have been used as shell materials to prevent oxidation [7].

The magnetic properties of nanoparticles are determined by many factors, mainly the nanoparticle size, shape or morphology, composition, and structure [2], these factors can be controlled during the synthesis of nanoparticles. There are various methods to prepare magnetite nanoparticles of nanometer size range. However, co-precipitation method is probably the simplest and most efficient synthetic route to obtain magnetic particles [8]. In the present work, magnetic nanoparticles were synthesized in aqueous solutions via co-precipitation of ferrous ($Fe^{2+}$) and ferric ($Fe^{3+}$) ions with the adding polyvinylpyrrolidone (PVP) to avoid the agglomeration and obtain nanoparticles with homogeneous size. The crystalline structure of the magnetite nanoparticles was characterized by aberration-corrected scanning transmission electron microscopy (STEM) and X-ray diffraction (XRD). Comparisons between the theoretical and experimental HAADF-STEM images are carried out and some insights from the atomic structure of the magnetic nanoparticles are discussed.

## 17.2 Experimental

For magnetite nanoparticles synthesis, 1M ferric chloride hexahydrate ($FeCl_3{\cdot}6H_2O$>99 %) and 1M ferrous chloride tetrahydrate ($FeCl_2{\cdot}4H_2O$>99 %) were prepared by dissolving iron salts in deionized water and after that hydrochloric acid (HCl) was added. In a typical experimental procedure, 10 ml of 1M $FeCl_3$ solution was mixed with 5 ml of 1M $FeCl_2$ solution in a flask. This solution was stirred, followed by the slow addition of 9 ml of 3M tetramethylammonium hydroxide pentahydrate ($C_4H_{13}NO{\cdot}5H_2O$), next 7 ml of 1M sodium hydroxide (NaOH) solution was added and finally, enough 1M of sodium borohydride ($NaBH_4$) solution was used to reach a pH of 13. At the end of the process, PVP solution was added. Vigorous stirring was continued for 20 min. The solution color changes from orange to black, leading to a black precipitate.

Structure of the magnetic nanoparticles was analyzed using an X-ray diffractometer (Siemens D5000, using Cu-Kα radiation). Measurements were made for 2θ values over 10–100°.

To characterize the magnetite nanoparticles by electron microscopy, copper grids with carbon film were prepared with a drop of the solution obtained. The samples were analyzed using a spherical aberration corrector for a probe-forming system in a Jeol ARM200F (200 keV) FEG-TEM/STEM. High-resolution high-angle annular dark field (HAADF)-STEM images were obtained. The probe current used in STEM mode was 23.2 pA using a condenser lens aperture size of 40 μm. The HAADF-STEM images were registered using a camera length of 80 mm and a collection angle of 50–180 mrad. HAADF-STEM image simulations have been performed using the QSTEM software package which uses the multislice algorithm. The parameters considered for the simulation correspond to the experimental conditions of the microscope.

## 17.3 Results and Discussion

The structure of the magnetic particles for drug carrier systems significantly affects the physicochemical properties, such as colloidal stability, drug release behavior, and magnetic intensity of the carrier; therefore, it is important to analyze the structure of the magnetic nanoparticles which will be used in the clinical applications [9]. Magnetite ($Fe_3O_4$) is a common magnetic iron oxide that has a cubic inverse spinel structure with oxygen atoms forming a face-centered cubic (fcc) closed packing and Fe cations occupying interstitial tetrahedral sites and octahedral sites. Figure 17.1 shows the XRD patterns of $Fe_3O_4$ nanoparticles synthesized by the co-precipitation method at different temperatures (25, 40, 80, and 90 °C). From the XRD patterns, the different peaks found could be well indexed to the cubic structure of $Fe_3O_4$ (JCPDS card no. 19-0629). However, some diffraction peaks from impurities were also observed. The average crystallite size at the different temperatures was obtained from the Scherrer equation, using the (311) crystalline plane of the cubic $Fe_3O_4$ structure, the values obtained were 20, 14, 18, and 18 nm at 25, 40, 80, and 90 °C, respectively. However, when the temperature increases, there are two reflections at around 25° on the XRD patterns of the samples that could be related with the presence of $Fe_2O_3$, by this reason a temperature synthesis less than 80 °C was used. Magnetite is thermodynamically unstable with respect to hematite ($Fe_2O_3$) in the presence of oxygen [10], even at room temperature.

Nowadays, the best technique to give information about the structure and composition of the nanomaterials is STEM. This technique offers significant benefits in

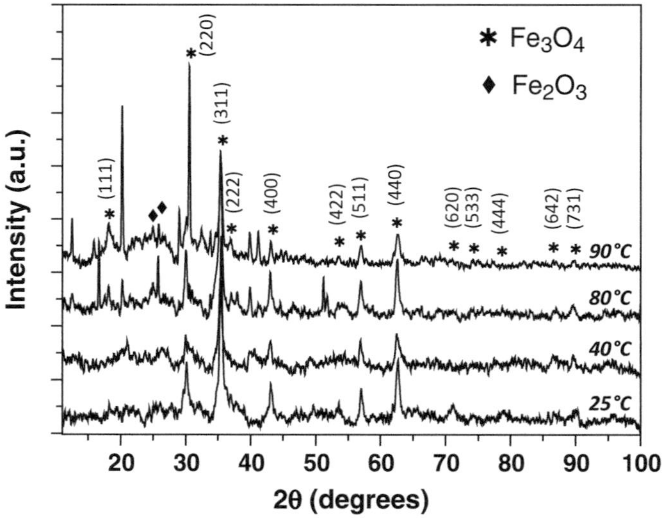


**Fig. 17.1** XRD patterns of $Fe_3O_4$ magnetic nanoparticles at different synthesis temperatures

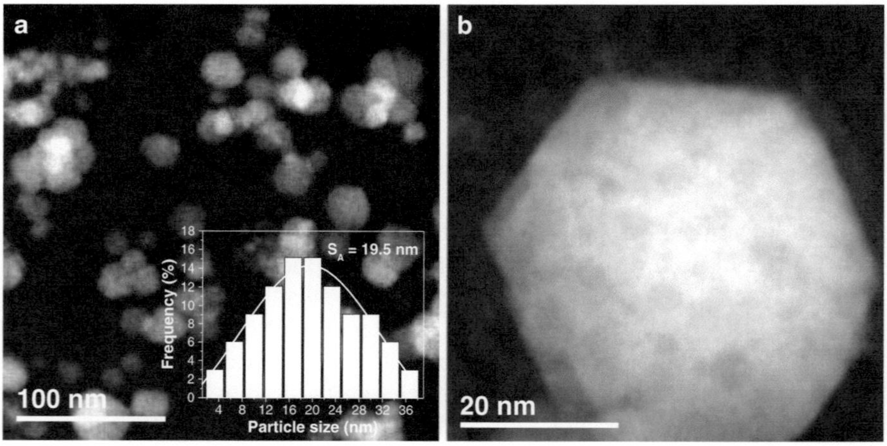


**Fig. 17.2** (**a**) HAADF-STEM image and particle size histogram of $Fe_3O_4$ magnetic nanoparticles. (**b**) Nanoparticle shows faceted surface planes

dark field operation with the HAADF imaging mode. The HAADF-STEM signal is directly proportional to the density and thickness of the specimen and proportional to $Z^{3/2}$ where Z is the atomic number. Therefore, it is possible to produce images with a contrast due to the mass-thickness or where the signal is proportional to the atomic number of the sample (Z-contrast). Figure 17.2a shows a HAADF-STEM image with its particle size histogram of the $Fe_3O_4$ nanoparticles. As can be observed the average particle size was 19.5 nm and the particles show faceted surface planes. Although the nanoparticles were dispersed, a strong agglomeration tendency due to

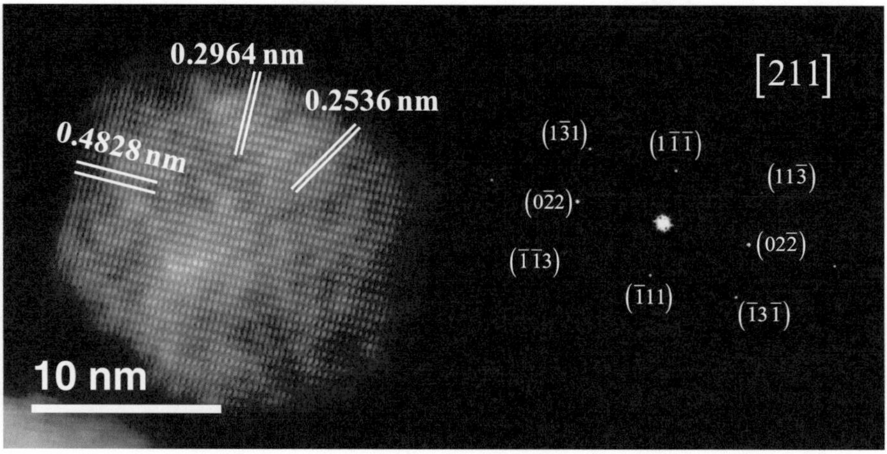


**Fig. 17.3** HAADF-STEM image with its FFT of $Fe_3O_4$ nanoparticle. The distances were successfully indexed to cubic-$Fe_3O_4$ in the [211] orientation

interparticle magnetic interactions is shown. Magnetic nanoparticles tend to agglomerate because of strong magnetic dipole–dipole attractions between particles combined with van der Waals force and high surface energy [11]. It is important to mention that almost all the nanoparticles show a hexagonal shape, which is associated with a truncated octahedron. Figure 17.2b shows a nanoparticle viewed close to [111] zone axis, which exhibit a hexagonal projection, the angle between the facets is 120° approximately and the measurement of the fringes spacing is 0.296 nm which is associated with the (220) crystalline plane of the fcc-$Fe_3O_4$ structure.

The successful implementation of aberration-correction in the STEM enables the discrimination of two objects separated by less than 1 Å, so that STEM is recognized as the most powerful tool to allow a single atom to be identified spectroscopically within a single atomic column of crystalline materials [12]. Figure 17.3 shows a high-resolution HAADF-STEM image of $Fe_3O_4$ nanoparticle. The measurement of the fringes spacing 0.4828, 0.2536, and 0.2964 nm corresponds to (1-1-1), (11-3), and (02-2) crystalline planes, respectively, and the nanoparticle is oriented in the [211] direction. Fast Fourier transform (FFT) shows the main reflections confirming the [211] zone axis. The experimentally and theoretically values calculated of the crystalline planes of the cubic-$Fe_3O_4$ structure show good agreement. The high-resolution HAADF-STEM image exhibits a well-defined lattice, in others words, straight fringes through the whole particle without stacking faults or twins, therefore could be assumed that the $Fe_3O_4$ nanoparticle is a single crystal [13].

Based on the fact that the contrast of the incoherent HAADF-STEM images strongly depends on the atomic number (Z-contrast), simulated HAADF-STEM images were also obtained. QSTEM program was used for computation of the incoherent HAADF-STEM images, which is based in the multislice method. This method evaluates the wave field in the crystal induced by the focused electron probe, thus, a complete integration of scattering intensities for each probe position

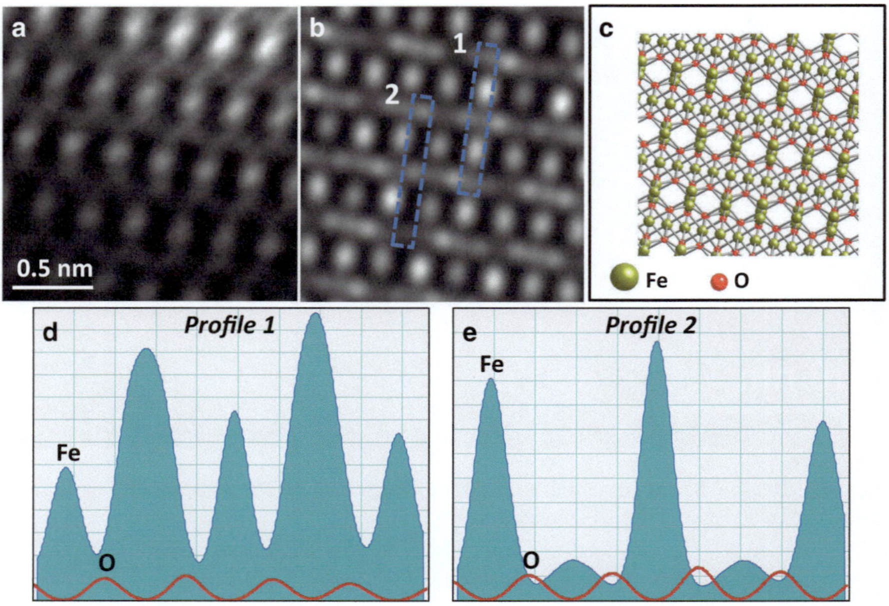


**Fig. 17.4** (**a**) High-resolution HAADF-STEM image of $Fe_3O_4$ nanoparticle, (**b**) simulated HAADF-STEM image of $Fe_3O_4$ nanoparticle, (**c**) model of the cubic structure of the magnetite, (**d**) and (**e**) profile intensity of the two marked regions in the simulated image

over a HAADF detector range is performed. Figure 17.4a shows a high-resolution HAADF-STEM image of $Fe_3O_4$ nanoparticle oriented in the [211] zone axis. As can be observed, the atomic resolution is clearly well defined, also it can be observed bright dots and an interface among them; however, at this point it is difficult to identify the positions of iron (Fe) or oxygen (O). For this reason, simulated HAADF-STEM images may help to understand the detailed structure in the image and how the image is formed and generated [14]. Figure 17.4b shows the simulated HAADF-STEM image of the $Fe_3O_4$ nanoparticle, as can be observed, the contrast of the experimental and simulated images is similar, therefore it can be concluded that the model is correct (Fig. 17.4c). For this, a supercell was performed from the $Fe_3O_4$ structure in the QSTEM program. In fact the contrast observed in the figures is mainly the scattering contribution of the Fe atoms; given that, the Fe has an atomic number higher than the O, 26 and 8, respectively, so that the scattering contribution of the O atoms is very weak. With respect to the interface, the contrast observed is the result of the scattering of the Fe atoms too. To verify this, profiles intensity from two regions shown in the simulated image were carried out. To obtain the profile intensity, first, the structure of $Fe_3O_4$ without the O atoms was simulated in order to obtain only the contribution of the Fe atoms. Similarly, at the structure of $Fe_3O_4$, the Fe atoms were removed to obtain only the contribution of the O atoms. At the end, the profiles were added. Figure 17.4d, e shows the two profiles intensity where the filled area corresponds to the Fe contribution and the red line to the oxygen contribution.

It can be seen that the oxygen contribution is very weak; however, the oxygen atoms are found on both sides of the iron columns, this is due to the orientation of the magnetic nanoparticle. The detection of light atoms with extremely weak scattering has been reported by using annular bright-field (ABF)-STEM [15, 16]. Usually, the ABF-STEM configuration is realized by covering the center region of the bright-field STEM detector. This technique was not performed in this work, however it is recommended to verify the O positions.

## 17.4 Conclusions

$Fe_3O_4$ magnetic nanoparticles were prepared through the co-precipitation method. The cubic structure of the magnetite $Fe_3O_4$ was determined by XRD and aberration-corrected (Cs) STEM. The average particle size was 19.5 nm. Some magnetic nanoparticles were found in [211] zone axis. The contrast of the HAADF-STEM images indicates the contribution mainly of the Fe atoms given that the contribution of the O atoms is very weak. This was checked by simulated HAADF-STEM images. $Fe_3O_4$ magnetic nanoparticles could be used as a potentially good magnetic support in magnetic technology, with good magnetic quality, and also they could have good biocompatibility.

**Acknowledgments** The authors acknowledge the Laboratorio Avanzado de Nanoscopia Electrónica "LANE" at the CINVESTAV-Zacatenco. Thanks to CONACyT because of the Becerril-Juárez's postdoctoral scholarship.

## References

1. Ghandoor HE, Zidan HM, Khalil MMH, Ismail MIM (2012) Synthesis and some physical properties of magnetite ($Fe_3O_4$) nanoparticles. Int J Electrochem Sci 7:5734–5745
2. Gubin SP, Koksharov YA, Khomutov GB, Yurkov GY (2005) Magnetic nanoparticles: preparation, structure and properties. Russ Chem Rev 74:489–520
3. Perez JM, O'Loughin T, Simeone FJ, Weissleder R, Josephson L (2002) DNA-based magnetic nanoparticle assembly acts as a magnetic relaxation nanoswitch allowing screening of DNA-cleaving agents. J Am Chem Soc 124:2856–2857
4. Sun J, Zhou S, Hou P, Yang Y, Weng J, Li X et al (2007) Synthesis and characterization of biocompatible $Fe_3O_4$ nanoparticles. J Biomed Mater Res A 80:333–341
5. Long R, Zhou S, Wiley BJ, Xiong Y (2014) Oxidative etching for controlled synthesis of metal nanocrystals: atomic addition and subtraction. Chem Soc Rev 43(17):6288–6310
6. Xia T, Wang J, Wu C, Meng F, Shi Z, Lian J, Feng J, Meng J (2012) Novel complex-coprecipitation route to form high quality triethanolamine-coated $Fe_3O_4$ nanocrystals: their high saturation magnetizations and excellent water treatment properties. CrystEngComm 14:5741–5744
7. Wei S, Wang Q, Zhu J, Sun L, Lin H, Guo Z (2011) Multifunctional composite core–shell nanoparticles. Nanoscale 3:4474–4502
8. Li XM, Xu G, Liu Y, He T (2011) Magnetic $Fe_3O_4$ nanoparticles: synthesis and application in water treatment. Nanosci Nanotechnol Asia 1:14–24

9. Fang C, Zhang MQ (2009) Multifunctional magnetic nanoparticles for medical imaging applications. J Mater Chem 19:6258–6266
10. Tang J, Myers M, Bosnick KA, Brus LE (2003) Magnetite $Fe_3O_4$ nanocrystals: spectroscopic observation of aqueous oxidation kinetics. J Phys Chem B 107:7501–7506
11. Lalatonne Y, Richardi J, Pileni MP (2004) Van der Waals versus dipolar forces controlling mesoscopic organizations of magnetic nanocrystals. Nat Mater 3:121–125
12. Pennycook SJ, Varela M, Hetherington CJD, Kirkland AI (2006) Materials advances through aberration-corrected electron microscopy. MRS Bull 31:36–43
13. Zhang L, Wu J, Liao H, Hou Y, Gao S (2009) Octahedral $Fe_3O_4$ nanoparticles and their assembled structures. Chem Commun 29:4378–4380
14. Kirkland EJ (2010) Advanced computing in electron microscopy, Second Edition, Springer Science+Business Media, LLC, New York
15. Ishikawa R, Okunishi E, Sawada H, Kondo Y, Hosokawa F, Abe E (2011) Direct imaging of hydrogen-atom columns in a crystal by annular bright-field electron microscopy. Nat Mater 10:278–281
16. Okunishi E, Ishikawa I, Sawada H, Hosokawa F, Hori M, Kondo Y (2009) Visualization of light elements at ultrahigh resolution by STEM annular bright field microscopy. Microsc Microanal 15:164–165

# Part VII
# Characterization of Materials for Industrial Applications

This part includes two chapters related to the characterization of materials for industrial application. The first chapter reported the growth and characterization of ZnS thin films on GaAs substrates, deposited at various temperatures (180–630 °C) using a radio frequency (RF) planar magnetron sputtering system. The direct growth of ZnS on GaAs substrates is advantageous because of not only it reduces the polar–nonpolar interface problems but also the thermal expansion coefficients are matched. Effects of temperature during the sputtering with Ar-plasma on crystalline quality, particle size of the thin films were studied. This research explores the possibility of using ZnS films in electroluminescent devices. Thus, the structural and optical properties of ZnS films have been studied in this part.

The second chapter deals with a green method for graphite exfoliation using a mechanochemical route. This method is low cost and environmentally friendly and avoids the use of highly corrosive and dangerous reagents. Natural graphite as raw material was used. The procedure consisted of a mixture of an equiatomic mix of graphite and sodium carbonate that is processed in a high-energy ball mill. Morphological and chemical characterization exhibits a notable reduction of particle size with an important increased level of defoliation. Some years ago common applications for exfoliated graphite were: seals, thermal insulators, resin composites, electrodes, lubricant supports, battlefield obscurants, molds, chemical reagents, and adsorption substrates. Actually, emerging fields are: high-speed transistors, transparent conducting films, batteries, structural nanocomposites, optical devices, and supercapacitors.

The main techniques used in the characterization of these materials are:

- Radio frequency (RF).
- Energy-dispersive X-ray spectroscopy (EDS).
- Photoluminescence (PL).
- X-ray diffraction (XRD).
- Scanning electron microscopy (SEM).
- High resolution transmission electron microscopy (HRTEM).
- Raman spectroscopy.

# Chapter 18
# A Green Method for Graphite Exfoliation Using a Mechanochemical Route

I. Estrada-Guel, F.C. Robles-Hernandez, and R. Martínez-Sánchez

**Abstract** In the present work, we proposed a method to manufacture exfoliated graphene. This method is low cost and environmentally friendly by mechanochemical means and avoids the use of highly corrosive and dangerous reagents. Our raw material consists of natural graphite and sodium carbonate. The mix is milled followed by a citric acid treatment used as defoliation agents. The mixture consists of an equiatomic mix of graphite and sodium carbonate that is processed in a high-energy ball mill. The milled and raw mixtures are leached with an aqueous acid solution, hot refluxed, washed, and overnight dried. Observed morphological and chemical evidence exhibits a notable reduction of particle size with an important increased level of defoliation that improved the surface area values.

**Keywords** Mechanochemical • Exfoliation • Graphite • Equiatomic • Natural

## 18.1 Introduction

Thirty years ago, the carbon family was limited to the natural graphite (Gr) and diamond members. A sudden change comes out with the discovery of molecular carbon allotropes as: fullerenes, carbon nanotubes and more recently, 2-D graphenes [1]. Graphenes are a single layer to less than ten stacked honeycomb carbon sheets [2] with $sp^2$-bonding. The carbon atoms are arranged in a densely packed honeycomb lattice [3] joined with covalent bonds linked in the perpendicular stacking direction by weak Van der Wall's interaction. The excellent mechanical, electrical, thermal, optical properties and possibility of manipulation via chemical functionalization

I. Estrada-Guel (✉) • R. Martínez-Sánchez
Centro de Investigación en Materiales Avanzados, CIMAV, Miguel de Cervantes 120, Chihuahua 31109, México
e-mail: ivanovich.estrada@cimav.edu.mx; roberto.martinez@cimav.edu.mx

F.C. Robles-Hernandez
Department of Mechanical Engineering Technology, University of Houston, Houston, TX 77204-4020, USA
e-mail: fcrobles@central.uh.edu

 
R. Pérez Campos et al. (eds.), *Materials Characterization*,
DOI 10.1007/978-3-319-15204-2_18

and surface modification make this material a growing field in the material science [4] and industry [1]. In the 1990s common applications for exfoliated graphite were: seals, thermal insulators, resin composites, electrodes, lubricant supports, battlefield obscurants, molds, chemical reagents, and adsorption substrates [5, 6]. Now, emerging fields are: high-speed transistors, transparent conducting films, batteries [7], structural nanocomposites, optical devices [8, 9], and supercapacitors. Unfortunately, graphene is sensitive to the content of defects and functional groups and affects the performance of these devices [10]. Hence, the development of methods to prepare graphene without added functional groups is an important issue for industrial applications [5]. There exist a variety of methods for synthesizing graphenes: mechanical [3] and electrochemical exfoliation [4, 6], micromechanical cleavage [2], microwave irradiation [11, 12], reduction of graphite oxide [5], graphite expansion [13, 14], liquid-phase exfoliation [15–17], etc. Some factors that determine the characteristics of graphene are: precursor morphology, particle size, and purity. Also, the used method and reagents determine the intercalation efficiency and stacking layers [18]. Graphene can be produced by physical and chemical exfoliation of bulk graphite using cost-efficient methods combined with other chemical treatments from inexpensive Gr [19]. Generally, exfoliation occurs when the graphene layers are forced apart by abrupt vaporization or decomposition of the intercalated species [19]. For successful exfoliation, overcoming the Van der Waals attractions between the adjacent layers is necessary [3]. Even though mechanical methods had demonstrated to be effective to exfoliate stacked materials [4, 20], chemical methods are preferred, its major disadvantage is the use of concentrated sulfuric acid or sulfur trioxide ($SO_3$) [9] and strong oxidizing agents as: nitric acid, potassium permanganate, and hydrogen peroxide [4, 16, 21], where the discharge of acid waste and its treatment have been pointed out to be troublesome [9]. This opens a window of opportunity for methods that are environmentally friendly, such as the one proposed herein. Here we present a methodology that consists in the exfoliation of graphite using a mechanochemical route, which avoids the use of corrosive and dangerous chemicals using sodium carbonate that is odorless and slightly alkaline along with citric acid that is a weak organic acid. Both chemicals are non-toxic and well known for their common use in the food industry.

## 18.2 Experimental

Raw materials are: natural graphite, sodium carbonate ($Na_2CO_3$), and citric acid ($H_3Cit$). An equiatomic mixture of $Na_2CO_3$ and Gr is milled in a SPEX 8000M high-energy mill using a hardened steel container using a ball-sample ratio of 5:1. The milling time is set to 0, 1, 2, 4, 8, and 16 h. Milled mixtures are leached with an aqueous solution of $H_3Cit$ 8 % (wt./wt.), hot refluxed for 2 h and washed with distilled water until neutral pH followed by a drying process at 80 °C overnight. X-ray diffraction analysis was carried out using a Panalytical X'pert pro diffractometer under the following conditions: 10°–110°, 0.0167° step and 60 s/step. Morphological and chemical studies of milled and lixiviated graphite powders were performed with scanning electron microscopy (SEM) with a JSM-7201F and transmission electron microscopy (TEM) in a JEM-2200FS coupled with an EDS Oxford Inst. Inca X-sight Mod. 7678. Raman spectrometry in a Horiba model Lab Ram HR, He-Ne laser 632.8 nm ± 1 $cm^{-1}$. Specific surface area was obtained by the Brunauer–Emmett–Teller (BET) equation using $N_2$.

## 18.3 Results and Discussion

### *18.3.1 Microstructural Characterization*

Figure 18.1 shows some micrographs (100X) of Gr/$Na_2CO_3$ mixtures at different milling times, where bright zones correspond to rich $Na_2CO_3$ areas while the black areas are composed by graphite. As a direct effect of milling, the components have been dispersed, obtaining a homogeneous mixture as the images shown. After citric acid treatment (leaching), $Na_2CO_3$ is eliminated, in form of sodium citrate in the aqueous washing solution, this compound is an innocuous substance commonly used as food antioxidant. This method in comparison is less hazardous than those previously reported [5, 8, 9].

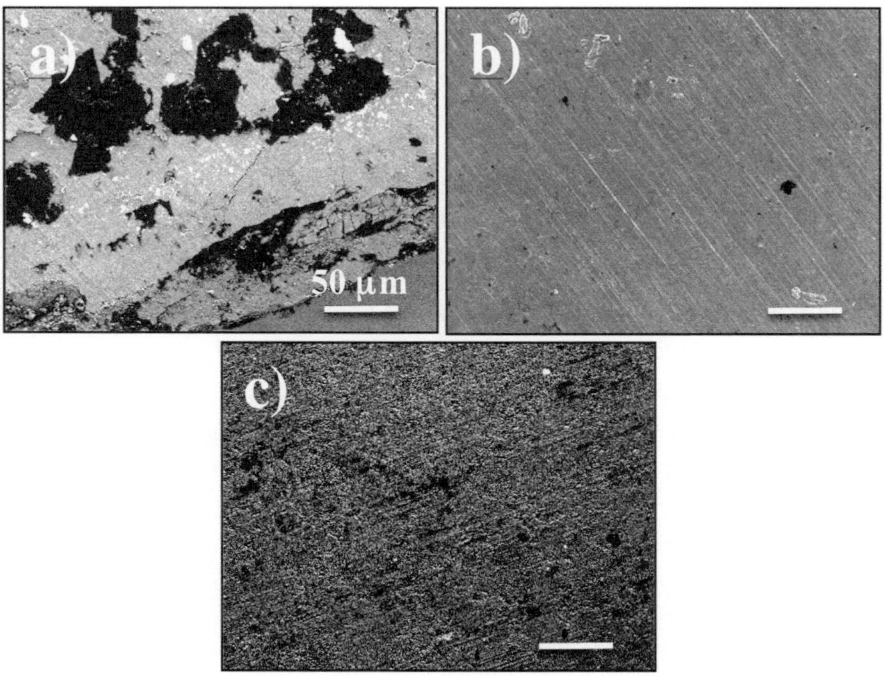


**Fig. 18.1** SEM micrographs of milled and compressed samples (compactation was carried out to obtain coarse specimens for more representative chemical analyses): (**a**) 0, (**b**) 1, and (**c**) 4 h

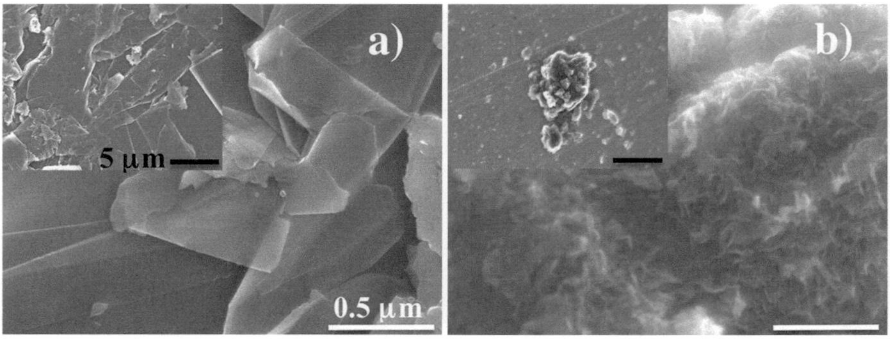


**Fig. 18.2** SEM micrographs of isolated Gr particles of (**a**) raw and (**b**) 8 h milled samples

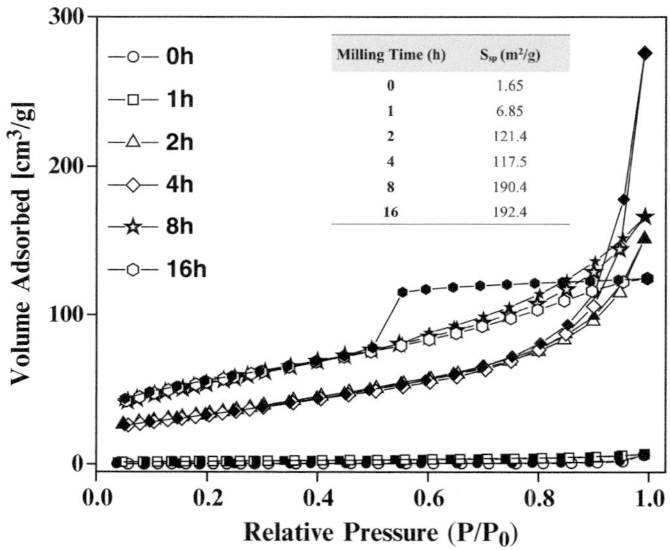


**Fig. 18.3** Isotherms of processed Gr samples and surface area determination

In Fig. 18.2 it is evident a notable reduction of particle size of samples after the mechanochemical process (left upper images), ordinary graphite structure was transformed by graphite layers sliding due to the impact forces released by ball collisions and an increased level of defoliation is noticed.

## *18.3.2 Surface Area Determination*

Adsorption/desorption isotherms showed that the properties in milled graphite depend on pore characteristics and specific surface area ($S_{sp}$), both are affected by milling. The inner table presented in Fig. 18.3 is an evidence of the accentuated increase in $S_{sp}$ as a function of milling time. The maximum surface area is found on those samples milled for 16 h.

The raw sample had a BET surface area of less than 2 that increases to 192 $m^2/g$ after 16 h of milling. From those results, it is clear that the milled samples show larger BET surface than the raw material that is in agreement with previous data of graphenes [1, 7]. In contrast, the surface area of the milled samples increases rapidly with milling times. The isothermals for the samples milled from 0 to 8 h are classified into an II type, according to BDDT/IUPAC classification, suggesting the existence of macropores, low hysteresis between adsorption and desorption isotherms implies a low proportion of mesopores. In contrast, 16 h sample presents hysteresis loop been classified as IV type. In this sample, we have a significant concentration of mesopores as reported by Makotchenko et al. [22]. Macropores can be more useful for the absorption of oil with more viscous and large molecular weight [9].

### 18.3.3 X-Rays Diffraction

In Fig. 18.4 are presented the X-rays diffraction (XRD) diffraction patterns of leached samples. From this figure is noticeable a marked reduction in the intensity, widening of the peaks along with the lost of secondary peaks for the sample milled for 16 h. This is a consequence of reduction of crystallinity of the samples as a direct result of processing. The full width at half maximum (FWHM) of main peak increases from 0.268 to 2.261 in samples milled for 0 and 16 h, respectively. It is important to mention that considerable intensity loss in main peak can reveal that the long-range periodicity associated with the staking of graphene in the c-axis in graphite is damaged [6].

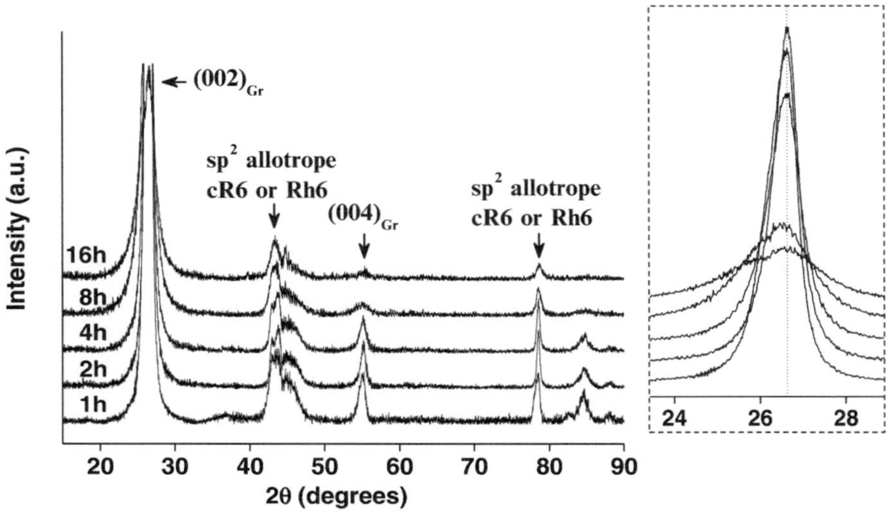


**Fig. 18.4** XRD spectra of leached Gr samples and magnification of main diffraction peak

Using the reflection at 26.56° is calculated the average interlayer spacing 0.338 ± 0.01 and 0.339 ± 0.01 nm for the 1 and 16 h samples respectively, this value is high compared to natural graphite flakes of 0.335 nm [18]. This can be related to a very slight separation between layers of graphenes in the graphite structure, without significant intercalation of new molecules or generation of oxygen functionality in the interlayer spacing of graphite [1], as can be seen in the right square of Fig. 18.4. Using the expression (18.1), the grain size of the investigated samples is calculated, where $D_p$ is the grain size, $\lambda$ is the wavelength of CuK$\alpha$ radiation (0.154056), $\beta$ is the FWHM, and $\theta$ is the Bragg's angle [23].

$$D_p = \frac{0.94 \cdot \lambda}{\beta \cos\theta} \tag{18.1}$$

The results indicate that the grain size in our materials changes from 31.83 to 3.77 nm from the raw material to the samples milled for 16 h with a corresponding lattice strain changes from 0.005 to 0.0418. The FWHM values follow the same path, except for the sample milled for 2 h that is attributed to a systematic error. From those results we can conclude that milling time helps refining the grain size, but it also induces residual stresses within the lattice. Here we identify the presence of the reflections around 40–45; $2\theta$ degrees. Those reflections were previously reported as diamond peaks [24–27], and more recently are attributed to a new carbon allotrope [28]. This allotrope has sixfold helical chains forming pure $sp^2$ arrangements. Based on our previous results this carbon allotrope has outstanding mechanical properties and is quite effective as reinforcement for structural composites. Based on the reported phase in [28] we identify the reflections in the Fig. 18.4 as cR6 and/or Rh6.

### 18.3.4 Raman Spectroscopy

The spectrum is mainly dominated for the typical D, G, and 2D bands. The presence in the D band is an indicator of small grain sizes and the 2D band clearly shows the presence of exfoliated layers that can be observed in Fig. 18.5.

Therefore, we assume that there is a presence of graphitic layers in all the samples; it is for this reason that we are presenting in the figure the ratios $I_G/I_D$ that had been used to determine the average lateral size of graphitic particles. This calculation is conducted using the expression (18.2) in nm, where $L_{las} = 638$ nm is the excitation laser wavelength, and $I_G$ and $I_D$ are the Raman intensity of the D and G bands, respectively [29]. The calculations indicate that the graphene layers range from 107 to 425.5 nm for the samples milled 1 and 4 h, respectively. The fact that samples milled for 16 h have larger particle sizes is an indication that milling is welding and fracturing the graphite constantly that is a well-known effect on milling. The D band is characteristic of defects and disorders and the D′ shoulder band is related to the disorder of edge carbons. The G band is related to the vibration of $sp^2$ bonded, usually graphite and carbon atoms in a two-dimensional hexagonal lattice.

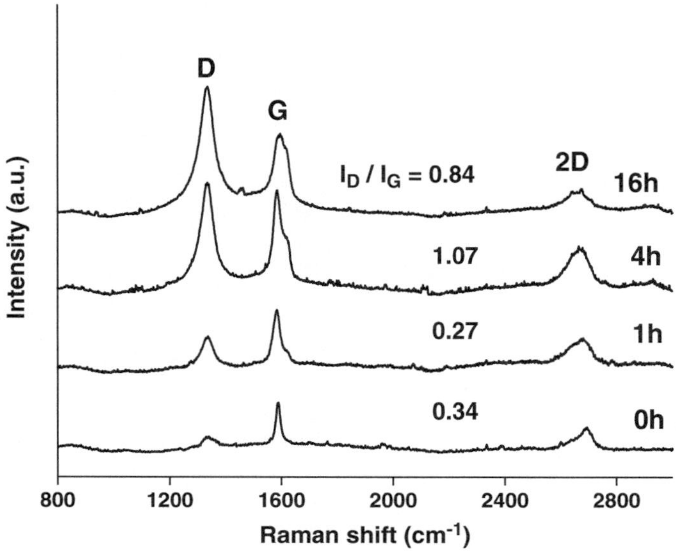


**Fig. 18.5** Raman spectra of leached graphite samples and D/G ratios

The intense 2D band originates from a two-phonon double resonance and is not seen in defect-rich graphene [1]. The shape, width, and position of the 2D peak are very sensitive to the number of graphene layers [18]. In single layer graphenes the 2D band is approximately two times stronger than the G band and it decreases with the number of layers. In our case we present in Fig. 18.5 a Raman spectra with lower rations that translate into multilayer graphene that also contributes to the disordered edge D band [11].

$$L_a = 2.4 \times 10^{-10} L_{las}^4 \cdot \frac{I_G}{I_D} \tag{18.2}$$

## 18.3.5 TEM and HRTEM

Figure 18.6 shows some micrograph of samples milled for 2 and 4 h and leached; in those images, we can observe the graphene layers on the bright field image (Fig. 18.6a) and the respective selected area diffraction pattern showing the typical pattern for graphene (Fig. 18.6b). The HRTEM image (Fig. 18.6c) shows the layered structures of graphitic carbon where the measured d spacing is 0.337 that corresponds to that on graphene and it is further confirmed with the results shown in Fig. 18.6d with a d spacing of 0.65 nm in the inset (FFT) and the d-spacing measured directly from the HRTEM image is 0.239 nm. Those values are comparable, but with marginal differences of less than 3 %, when comparable to the values

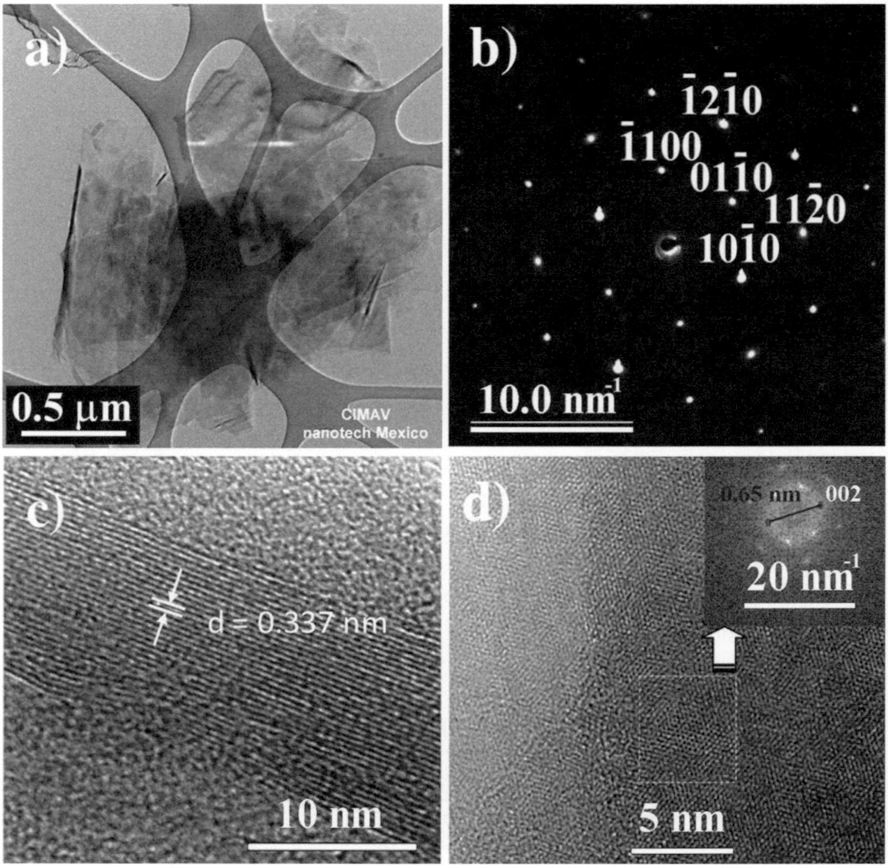


**Fig. 18.6** (**a**, **b**) TEM and (**c**, **d**) HRTEM images of processed samples: (**a**, **b**) 2 h and (**c**, **d**) 4 h. (**a**) Low magnification TEM image, (**b**) SADP in Z=[0001], (**c**) high magnification image of specimen showing its interplanar distance and (**d**) FFT image of milled sample

reported in [26, 30, 31]. In Fig. 18.6d is presented the typical honeycomb structure of graphite or graphene, the measured lattice parameter in this case is 0.201 nm. An EDS chemical analysis gives us a composition of 98.1 at.% C and 1.9 at.% O, being the oxygen presence common in this type of carbons. Therefore, the results showed in Fig. 18.6 confirm the graphene nature of the investigated material.

## 18.4 Conclusions

Here we demonstrated the effectiveness to exfoliation graphene from graphite using mechanical milling that is a green technology. The characterization methods used herein clearly confirm the presence of graphitic carbon and graphene in the milled products. The BET surface area increases exponentially with milling time to a maximum of 192 $m^2/g$ after 16 h of milling. We identify an allotrope of carbon (cR6 or hR6) that was previously reported as an effective reinforcement for composites. The corresponding grain size in the synthesized particles ranges from 107 to 425 nm in Raman for the raw and milled sample for 16 h. In XRD the same samples show sizes between 31.8 and 3.8 nm. The discrepancies between Raman and XRD are because Raman predicts the length of the crystal (breath mode) while XRD is used to assess the stacking. Further, the HRTEM work clearly confirms the presence of graphene in the milled samples with d-spacing approximately 3 % different from the theoretical.

**Acknowledgments** This research was supported by CONACYT (Project No. 169262) and the Redes Temáticas de Nanociencias y Nanotecnología (124886). The technical assistance of P. P. Ruiz, E. T. Moye, and J. C. Cisneros is gratefully recognized.

## References

1. Kuila T, Bose S, Mishra AK, Khanra P, Kim NH, Lee JH (2012) Chemical functionalization of graphene and its applications. Prog Mater Sci 57:1061–1105
2. Salvatierra RV, Domingues SH, Oliveira MM, Zarbin AJG (2013) Tri-layer graphene films produced by mechanochemical exfoliation of graphite. Carbon 57:410–415
3. Cai MZ, Thorpe D, Adamson DH, Schniepp HC (2012) Methods of graphite exfoliation. J Mater Chem 22:24992–25002
4. Rao KS, Senthilnathan J, Liu YF, Yoshimura M (2014) Role of peroxide ions in formation of graphene nanosheets by electrochemical exfoliation of graphite. Nat Sci Rep 4:4237–4242
5. Chung DDL (1987) Exfoliation of graphite. J Mater Sci 22:4190–4198
6. Chung DDL (1987) Intercalate vaporization during the exfoliation of graphite intercalated with bromine. Carbon 25:361–365
7. Badi N, Erra AR, Hernandez FC, Okonkwo AO, Hobosyan M, Martirosyan KS (2014) Low-cost carbon-silicon nanocomposite anodes for lithium ion batteries. Nanoscale Res Lett 9:360
8. Bao J, Xing S, Wang Y, Wu W, Robles-Hernandez F, Pei S-s (2013) Fabrication of large-area twisted bilayer graphene for high-speed ultra-sensitive tunable photodetectors. In: George T, Saif Islam M, Dutta AK (ed) Proceedings of SPIE 8725, micro- and nanotechnology sensors, systems, and applications V, p 872503
9. Wang Y, Su Z, Wu W, Nie S, Lu X, Wang H, McCarty K, Pei S-s, Robles-Hernandez F, Hadjiev VG (2013) Anomalously strong 2D band intensity in twisted bilayer graphene: Raman evidence for doubly degenerate dirac band. arXiv preprint arXiv:1309.5496
10. Zhou M, Tang J, Cheng Q, Xu GJ, Cui P, Qin LC (2013) Few-layer graphene obtained by electrochemical exfoliation of graphite cathode. Chem Phys Lett 572:61–65
11. Tryba B, Morawski AW, Inagaki M (2005) Preparation of exfoliated graphite by microwave irradiation. Carbon 43:2417–2419
12. Yan J, Fan ZJ, Wei T, Qian WZ, Zhang ML, Wei F (2009) Preparation of exfoliated graphite containing manganese oxides with high electrochemical capacitance by microwave irradiation. Carbon 47:3371–3374
13. Lee BJ (2002) Characteristics of exfoliated graphite prepared by intercalation of gaseous $SO_3$ into graphite. Bull Korean Chem Soc 23:1801–1805
14. Manning TJ, Mitchell M, Stach J, Vickers T (1999) Synthesis of exfoliated graphite from fluorinated graphite using an atmospheric-pressure argon plasma. Carbon 37:1159–1164
15. Bourlinos AB, Georgakilas V, Zboril R, Steriotis TA, Stubos AK (2009) Liquid-phase exfoliation of graphite towards solubilized graphenes. Small 5:1841–1845

16. Sim Y, Park J, Kim YJ, Seong MJ, Hong S (2011) Synthesis of graphene layers using graphite dispersion in aqueous surfactant solutions. J Korean Phys Soc 58:938–942
17. Song XJ, Shi Z, Tan XH, Zhang SH, Liu GS, Wu KB (2014) One-step solvent exfoliation of graphite to produce a highly-sensitive electrochemical sensor for tartrazine. Sens Actuators B Chem 197:104–108
18. Vieira F, Cisneros I, Sansiviero MTC, Miranda AM, Rosa NG, Lima UB, Mohallem NDS (2006) Preparation processes and properties of expanded graphite for alkaline batteries. J Phys Chem Solids 67:1208–1212
19. Falcao EHL, Blair RG, Mack JJ, Viculis LM, Kwon CW, Bendikov M, Kaner RB, Dunn BS, Wudl F (2007) Microwave exfoliation of a graphite intercalation compound. Carbon 45:1367–1369
20. Asghar HMA, Hussain SN, Sattar H, Brown NW, Roberts EPL (2014) Environmentally friendly preparation of exfoliated graphite. J Ind Eng Chem 20:1936–1941
21. Inagaki M, Tashiro R, Washino Y, Toyoda M (2004) Exfoliation process of graphite via intercalation compounds with sulfuric acid. J Phys Chem Solids 65:133–137
22. Makotchenko VG, Grayfer ED, Nazarov AS, Kim SJ, Fedorov VE (2011) The synthesis and properties of highly exfoliated graphites from fluorinated graphite intercalation compounds. Carbon 49:3233–3241
23. Drits V, Srodon J, Eberl DD (1997) XRD measurement of mean crystalline thickness of illite and illite/smectite: reappraisal of the Kubler index and the Scherrer equation. Clay Clay Miner 45:461–475
24. Fals AE, Hadjiev VG, Robles Hernández FC (2012) Multi-functional fullerene soot/alumina composites with improved toughness and electrical conductivity. Mater Sci Eng A 558:13–20
25. Fals AE, Hadjiev VG, Robles Hernández FC (2013) Porous media reinforced with carbon soots. Mater Chem Phys 140:651–658
26. Fals AE, Quintero J, Hernández FCR (2010) Manufacturing of hybrid composites and novel methods to synthesize carbon nanoparticles. In: MRS online proceedings library, vol 1276, null–null
27. Robles Hernández FC, Calderon HA (2012) Nanostructured Al/$Al_4C_3$ composites reinforced with graphite or fullerene and manufactured by mechanical milling and spark plasma sintering. Mater Chem Phys 132:815–822
28. Wang J-T, Chen C, Wang E, Kawazoe Y (2014) A new carbon allotrope with six-fold helical chains in all-$sp^2$ bonding networks. Sci Rep 4:4339
29. Cancado LG, Takai K, Enoki T, Endo M, Kim YA, Mizusaki H, Jorio A, Coelho LN, Magalhaes-Paniago R, Pimenta MA (2006) General equation for the determination of the crystallite size L-a of nanographite by Raman spectroscopy. Appl Phys Lett 88(163106):1–3
30. Baskin Y, Meyer L (1955) Lattice constants of graphite at low temperatures. Phys Rev 100:544
31. Ferrari AC, Robertson J (2000) Interpretation of Raman spectra of disordered and amorphous carbon. Phys Rev B 61:14095–14107

# Chapter 19
# Characterization of Wurtzite Type ZnS Grown by RF Magnetron Sputtering

**Joel Díaz-Reyes, Roberto S. Castillo-Ojeda, and Javier Martínez-Juárez**

**Abstract** ZnS was grown on (001) GaAs substrates at different temperatures by RF magnetron sputtering. ZnS chemical stoichiometry was determined by energy-dispersive X-ray spectroscopy (EDS). The XRD analysis and Raman scattering reveal that ZnS deposited thin films showed hexagonal crystalline phase wurtzite type. The average crystallite size range of the film was from 8.15 to 31.95 nm, which was determined using the Scherrer–Debye equation on diffractograms. Besides, an experimental study on first- and second-order Raman scattering of ZnS films is made. The observed emission peaks in the room temperature photoluminescence spectra are associated at oxygen traps and interstitial sulphur.

**Keywords** Wurtzite • RF magnetron sputtering • Energy-dispersive X-ray spectroscopy (EDS) • XRD • Crystalline

## 19.1 Introduction

ZnS is a semiconductor with direct wide band gap to room temperature which continues gaining ground in recent years in view of its application potential in optoelectronic devices such as light emitting diodes, flat panel displays, nonlinear optical devices, sensors, lasers and photocatalysis [1, 2]. Owing to its wide band

J. Díaz-Reyes (✉)
Centro de Investigación en Biotecnología Aplicada, Instituto Politécnico Nacional, Ex-Hacienda de San Juan Molino, Km. 1.5, Tepetitla, Tlaxcala 90700, México
e-mail: joel_diaz_reyes@hotmail.com

R.S. Castillo-Ojeda
Universidad Politécnica de Pachuca, Km. 20, Rancho Luna, Ex-Hacienda de Santa Bárbara, Municipio de Zempoala, Hidalgo 43830, México
e-mail: rcastillo_ojeda@yahoo.com.mx

J. Martínez-Juárez
Centro de Investigación en Dispositivos Semiconductores del Instituto de Ciencias de la Universidad Autónoma de Puebla, Benemérita Universidad Autónoma de Puebla, 14 Sur y San Claudio S/N, Ciudad Uniuversitaria, Edif. No. 137. Col. San Manuel, Puebla, Puebla 72570, México
e-mail: javmartinez11@gmail.com

 
R. Pérez Campos et al. (eds.), *Materials Characterization*,
DOI 10.1007/978-3-319-15204-2_19

gap, it is used in violet and blue regions. Nevertheless for some uses, it has been shown that such materials must be grown with a monocrystalline structure and a smooth surface [3]. ZnS is mostly found in one of two structural forms cubic sphalerite or hexagonal wurtzite, which have wide band gaps of 3.54 and 3.80 eV, respectively, at 300 K [4]. It is a well-known luminescent material having prominent and promising applications in displays, sensors and blue-light emission device application [5]. The main advantage of the nanostructures is their size dependent property. The Bohr radius of ZnS is 2.5 nm [6]. The recombination properties of the ZnS nanofilms differ from those of bulk ZnS due to the larger surface to volume ratio. With the reduction in size, the surface area plays a dominant role in the material properties, and the presence of many surface defects may also affect the emission properties of ZnS nanostructures.

In this work are reported the growth and characterization studies of ZnS thin films on (001) GaAs deposited at various temperatures (180–630 °C) using an rf planar magnetron sputtering system. The effects of temperature during the sputtering with Ar-plasma on crystalline quality and particle size of the thin films were studied by energy-dispersive X-ray spectroscopy (EDS), X-ray diffraction, Raman spectroscopy and room temperature photoluminescence. To explore the possibility of using it in electroluminescent devices, a study of the structural and optical properties of the host material is an important step. Based on the above criterion, the structural and optical properties of ZnS films have been studied in the present work.

## 19.2 Experimental Details

The used substrates were (001) GaAs semi-insulating doped with chromium of $10 \times 10$ mm$^2$ of area, with 50 mm separation between the target and substrate. The direct growth of ZnS on GaAs substrates is advantageous because of not only it reduces the polar–nonpolar interface problems but also the thermal expansion coefficients are matched. The base pressure inside the chamber was lower than $7.5 \times 10^{-7}$ mbar. Plasma of Ar (99.999 %) was created to sputter a 100 mm diameter, water-cooled, ZnS (99.99 %) target mounted under a planar magnetron. The RF power was 50 W. The sputtering time was about 2 h for all samples. During sputtering process the Ar-pressure was maintained at $15 \times 10^{-3}$ mbar. A partial pressure of hydrogen gas was introduced to the growth chamber during 30 min at 300 K before film deposition on the GaAs substrate. This gas partial pressure condition was maintained during the temperature ramp from 300 K to the growth temperature. The films crystalline phase and structure were determined with a Bruker D8 Discover diffractometer using the copper K$\alpha$ radiation ($\lambda = 1.5406$ Å) at 40 kV and 40 mA with parallel beam geometry. The ZnS stoichiometry was estimated by measurements of EDS that were carried out in an LEO 438VP system, with W.D. of 26 mm using a pressure of 20 Pa. Raman scattering experiments were performed at 300 K using the 6,328 Å line of a He-Ne laser at normal incidence for excitation. The nominal laser power used in these measurements was 20 mW. Scattered light was analysed using a micro-Raman system (Lambram model of Dilor), a holographic notch filter made by Kaiser Optical System, Inc. (model superNotch-Plus), and a 256×1024-pixel CCD used as detector cooled to 140 K using liquid nitrogen. Photoluminescence was taken with a solid state laser at 325 nm with 60 mW as excitation source and a Sciencetech 9040 monochromator was used to perform the sweep of wavelength at room and low temperature in a Cryogenics cryostat.

## 19.3 Results and Discussion

Figure 19.1 shows XRD diffractograms from the polycrystalline ZnS nanostructures grown on (001) GaAs. It is observed from XRD patterns that ZnS films deposited even at low temperature are in crystalline nature, which present two peaks dominant at 31.59° and 65.94°. This may be attributed to the fact that the crystalline GaAs substrate facilitates the growth of crystalline ZnS thin films and their crystal lattice orientation is initiated on the GaAs substrates at room temperature. The obtained structural parameters with the software DICVOL04 data are in good agreement with the published ones. From this close agreement, it is confirmed that deposited ZnS films for all the growth temperatures belong to the wurtzite crystal system. The X-ray pattern of deposited ZnS films is described in the $C_{6v}$ (6 mm) and whose lattice parameters were calculated using the software DICVOL04, obtaining the following lattice parameters values: $a=3.81$ Å and $c/a=1.62$, which are in agreement with the reported values [7]. All the samples present the same reflections and some of them can be assigned. There is a peak highly intense from the ZnS/GaAs nanostructures at 65.94°, which is composed by two peaks that correspond to the wurtzite hexagonal

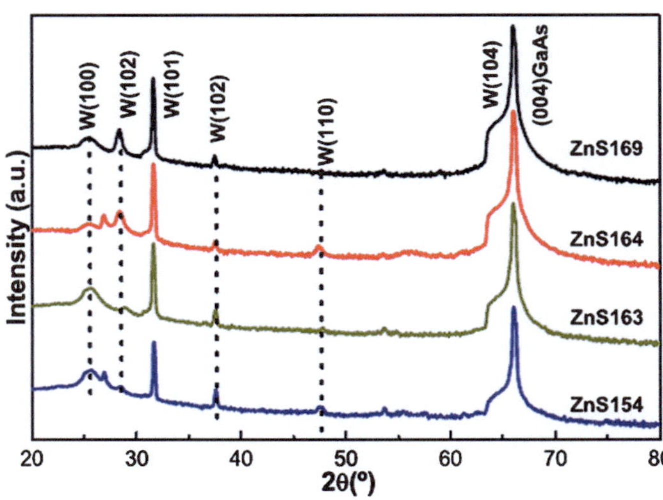


**Fig. 19.1** X-ray patterns of the ZnS nanofilms synthesized by RF magnetron sputtering

**Table 19.1** Results of the analysis by EDS of the samples grown by RF magnetron sputtering

| Sample | $T_G$ (°) | Sulphur atomic percentage (%) | Zinc atomic percentage (%) | Grain size by XRD (nm) |
|---|---|---|---|---|
| ZnS140 | 180 | 50.21 | 49.79 | 28.18 |
| ZnS154 | 270 | 49.01 | 50.99 | 31.95 |
| ZnS163 | 335 | 47.07 | 52.93 | 28.31 |
| ZnS164 | 450 | 38.89 | 61.11 | 30.30 |
| ZnS169 | 630 | 32.18 | 67.18 | 8.15 |

It presents the grain size of the ZnS nanoparticles calculated from hexagonal (101) reflection

(104) reflection of ZnS and zincblende (004) GaAs reflection that comes from the substrate material. As is observed from the X-ray diffractograms the intensity of the peaks clearly increases as the grown temperature is increased, in particular the peak at 65.94° of sample ZnS169 that is seven orders of magnitude higher than the peak of the sample ZnS154 as well as in increase of the peak at 28.45° associated with the hexagonal (002) reflection as can be seen in Fig. 19.1. Other peaks are observed on the X-ray diffractograms of the samples. The hexagonal (101) reflection is identified at 31.59°, hexagonal (002) reflection is observed at 28.45° and hexagonal (100) at 25.3°. It is clear that increasing the temperature improves the crystalline structure in nanofilms [8], since main peak increases as the grown temperature is increased, which implies that the crystallinity is better as previously mentioned. Some authors have reported that low growth temperature (<335 °C) the ZnS films result in a cubic structure for the thin films [9], but in this case did not occur and all nanostructures were wurtzite type. To obtain more structural information, the mean grain size of the deposited nanofilms was evaluated using the Scherrer equation. The mean grain size was calculated from hexagonal (101) reflection of ZnS for the samples. The mean grain sizes of the nanostructures are presented in Table 19.1. These results show that the grain size remained almost constant (~28.68 nm) for the temperature range from 180 to 450 °C, to higher temperatures the grain size diminishes as can be observed in Table 19.1. The chemical composition of the grown samples was determined by EDS, which furthermore allows us to know the presence of any unintentionally incorporated impurities. The EDS spectra indicate the presence of oxygen in the samples along with zinc and sulphur, they are not shown. The oxygen molar fractions in two ZnS154 and ZnS169 samples are 8.24 and 9.53 %, respectively. The X-ray photoelectron spectroscopy (XPS) analysis has showed that oxygen ions are present in the sample in the form of $O^{2-}$ ions that may produce ZnO.

Figure 19.2 illustrates the 300 K photoluminescence spectra of two intentionally undoped typical ZnS samples, ZnS154 and ZnS169, which were grown at lower and higher temperature, respectively. The Gaussian fit of the PL spectrum for the ZnS154 sample shows a strong violet peak at 2.95 eV, two strong blue emission bands ~2.85 and ~2.65 eV, and a green emission band at 2.47 eV. Figure 19.2b illustrates the Gaussian fit of the PL spectrum in the violet–blue–green–orange region for the ZnS169 sample, which shows a strong green emission at 2.32 with a shoulder at ~2.53 eV in the blue region, a deep level related orange emissions at ~2.10 eV

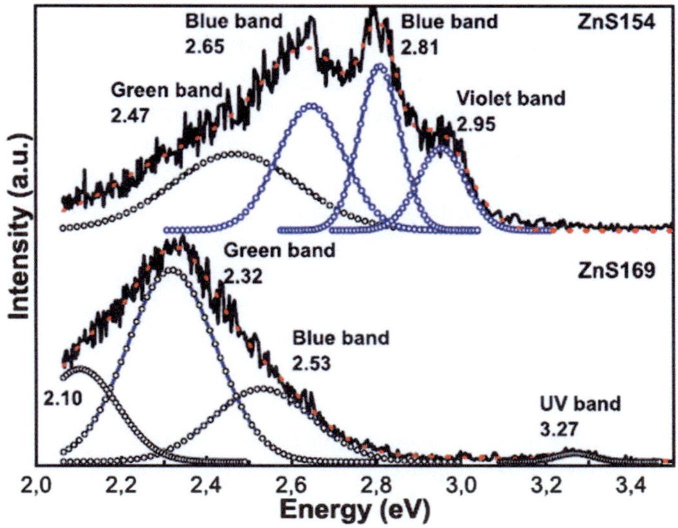


**Fig. 19.2** Room-temperature photoluminescence spectra of the samples: ZnS154 and ZnS169

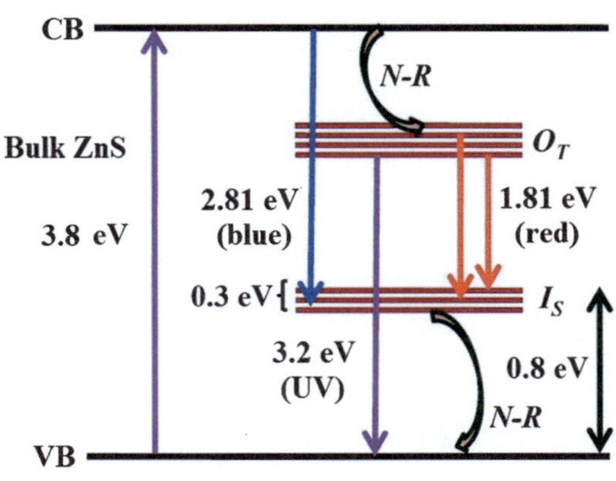


**Fig. 19.3** Energy level diagram explaining the observed emission bands. *VB* valence band, *CB* conduction band, *N-R* non-radiative transitions, *OT* oxygen trap levels, *IS* interstitial sulphur states

as well as a weak UV band at ~3.27 eV. All the radiative emission bands observed in the present samples can be explained on the basis of the energy level diagram shown in Fig. 19.3, which is an adaptation of the one reported by Goswami and Sen [10]. The analysis is done considering the presence of defect levels due to oxygen ions and interstitial sulphur ions in the system. The radiative transition from oxygen trap levels to valence band in ZnS is reported to be in the UV region around 381 nm whereas the transition from the conduction band to the defect levels of interstitial sulphur ions is reported in the blue region around 440 nm [11]. The interstitial S states marked as $I_S$ in figure can act as deep acceptor levels.

One can comment on the intensity enhancement of the 2.64–3.2 eV range and specifically to the formation of distinct peaks at 2.65, 2.81 and 2.95 eV. In ZnS, the valence band consists largely of *s* and *p* orbitals from sulphur while the conduction band is mainly due to *s*-orbitals of zinc [12]. Although little is known about the energy levels of interstitial sulphur or zinc in ZnS, studies of the green luminescence of CdS powder suggest that interstitial Cd and S vacancies act as "shallow donors" (electron traps) and interstitial S and Cd vacancies can behave as "deep acceptor" (hole traps). If a similar band structure applies to powdered ZnS, the blue emission can be understood as follows. Figure 19.3 illustrates the scheme of the energy states nanocrystalline ZnS band structure and the various processes involved during the PL of these materials. Sulphur vacancies in ZnS generate localized donor sites just below the conduction band. Excitation of these produces a positive charge and conduction band electrons. This localized charge exerts a potential, which can further trap electrons. So upon excitation, $S^{2-}$ vacancies are pumping the electrons into conduction band. Emission occurs when a captured electron recombines with a hole in the valence band or in some acceptor level, which are interstitial sulphur states in this case. These emissions are similar to deep transitions, because here $S^{2-}$ vacancies are behaving like impurity levels. Therefore, blue and violet emissions are attributed lattice defects related to sulphur and zinc vacancy or interstitials.

Figure 19.4 depicts the Raman spectra measured on the wurtzite-type ZnS films at 300 K. Wurtzite ZnS unit cell belongs to the $C_6$v point group symmetry with four atoms in a unit cell. The zone-centre optical phonons can be classified as the following irreducible representations: $\Gamma_{opt} = A_1 + E_1 + 2E_2 + 2B_1$. The $B_1$ modes are silent modes, $A_1$ and $E_1$ modes are polar modes and each splits into transverse-optical (TO) and longitudinal-optical (LO) components with different frequencies and both Raman and infrared active, and $E_2$ modes are nonpolar and Raman active only have

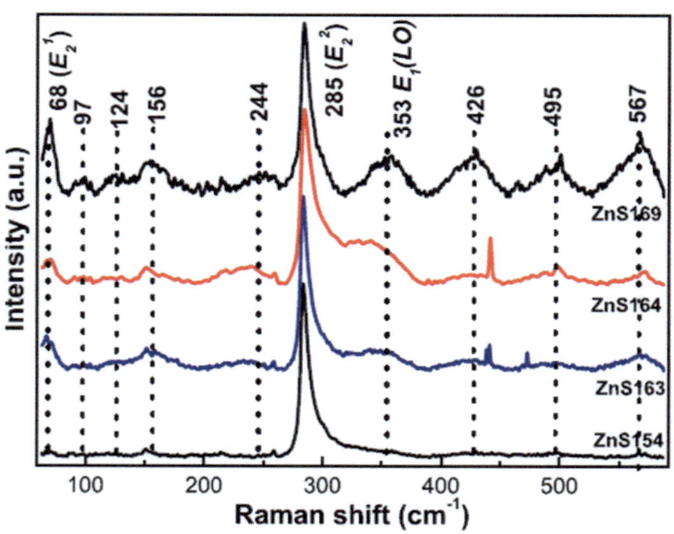


**Fig. 19.4** Raman scattered spectra at 300 K from ZnS layers for different grown temperatures

**Table 19.2** First-order Raman frequencies of theoretical and experimental results for hexagonal wurtzite ZnS

| Parameters | [13] | [14] | [15] | [13] | This work |
|---|---|---|---|---|---|
| $E_2^1$ | | 72 | 69 | 76 | 68 |
| $A_1$(TO) | 275 | 273 | 272 | 287 | 244 |
| $E_1$(TO) | 279 | 273 | 276 | 288 | – |
| $E_2^2$ | 285 | 286 | 286 | 296 | 285 |
| $E_1$(LO) | 353 | 351 | 351 | 347 | 350 |
| $A_1$(LO) | 353 | 351 | | 350 | – |

two frequencies, the higher-frequency $E_2$ (high) mode is associated with the sulphur atoms and the lower-frequency $E_2$ (low) mode is associated with the vibration of the heavy Zn sublattice. The reported frequencies of the Raman active modes of wurtzite ZnS are shown in Table 19.2. The Raman spectrum of sample ZnS154 mainly presents as dominant mode $E_2$, which corresponds to $E_2^2$ and is observed around 285 cm$^{-1}$ that confirms that the structure is wurtzite type [13]. As is observed the vibrational mode is dominant in this spectrum and is due to abundance of sulphur in the sample. As the growth temperature is increased the S molar fraction decreases appearing other peaks as $E_2^1$ at 68 cm$^{-1}$, which is associated with the vibration of the heavy Zn sublattice, as can be seen in Fig. 19.4.

The optical phonon mode $A_1$(LO) = $E_1$(LO) is reported to have the highest intensity in wurtzite ZnS structure, which was observed at 350 cm$^{-1}$. The relative intensity of this peak is a measure of the concentration of crystalline defects in the system. The intensity of the LO phonon mode is found to be comparable with that of the other optical modes in the present case. This is expected in view of the presence of defects in the system, as was inferred from luminescence studies. The LO phonon mode shows a red shift compared to the reported bulk value. Red shift of the LO phonon mode is known to occur in nanocrystalline systems due to spatial confinement as well as tensile stress due to the presence of defects. Since PL studies indicate the presence of defects in the form of oxygen trap ions and interstitial sulphur ions, the red shift observed here could be a combined effect of confinement as well as strain due to defects in the lattice.

Brafman and Mitra have reported that the modes $A_1$(TO) and $E_1$(TO) occur at the same wavenumber 273 cm$^{-1}$ [14]. However, Cheng et al. have calculated theoretically and later confirmed through experimental results that these modes can have different values [13]. The peak observed at 244 cm$^{-1}$ in the present work is attributed to $A_1$(TO) that is shifted to red. The composite peak near 166 cm$^{-1}$ can be due to the overlap of peaks at 158 and 177 cm$^{-1}$, attributed to two-phonon processes [14]. In the low-frequency region (Fig. 19.4), there are five weak peaks at 78, 107, 134, 166 and 254 cm$^{-1}$. The peak at 254 cm$^{-1}$ is assigned to LA overtones along M-K, where phonon dispersion is nearly flat and hence the DOS is very high. Above 380 cm$^{-1}$, we find the acoustic and optical combinations. For instance, the broad weak peak at 436 cm$^{-1}$ can be assigned to TA + LO at L point for flat phonon dispersion. The second-order Raman scattering in the high-frequency region corresponds to LO overtones and combinations involving LO modes [16].

## 19.4 Conclusions

The ZnS nanofilms grown by rf magnetron sputtering on GaAs (001) have been studied by XRD, EDS, room-temperature photoluminescence. XRD measurements show that the grain size remained almost constant (~28.68 nm) for the temperature range from 180 to 450 °C. XRD analysis reveals that the sample grown at 180 °C is polycrystalline and strongly textured, the samples from 270 to 450 °C are polycrystalline with hexagonal phases, and the orientation for sample 630 °C is believed to be influenced by the substrate orientation. ZnS/GaAs nanostructures thus produced that the samples exhibit different photoluminescent properties in the visible region as compared to their bulk counterpart.

## References

1. Tan M, Cai W, Zhang L (1997) Optical absorption of ZnS nanocrystals inside pores of silica. Appl Phys Lett 71:3697–3699
2. Yokogawa T, Taguch T, Fujita S, Satoh M (1983) Intense blue-emission band and the fabrication of blue light emitting diodes in I-doped and Ag-ion-implanted cubic ZnS. IEEE Trans Electron Devices 30:271–277
3. Theis D (1984) Selected analytical tools yield a better insight into electroluminescent thin films. Phys Status Solidi (a) 81:647–655
4. Fang J, Holloway PH, Yu JE, Jones KS, Pathangey B, Brettschneider E, Anderson TJ (1993) MOCVD growth of non-epitaxial and epitaxial ZnS thin films. Appl Surf Sci 70/71:701–706
5. Tran KT, Park W, Tong W, Kyi MM, Wagner KB, Summers JC (1997) Photoluminescence properties of ZnS epilayers. J Appl Phys 81:2803–2809
6. Kumbhojkar N, Nikesh VV, Kshirsagar A, Mahamuni S (2000) Photophysical properties of ZnS nanoclusters. J Appl Phys 88:6260–6264
7. Cook JW Jr, Eason BD, Vaudo PR, Schetzina FJ (1992) Molecular-beam epitaxy of ZnS using an elemental S source. J Vac Sci Technol B 10:901–904
8. Boudghene-Stambouli A, Hamzaoui S, Bouderbala M (1996) Blue emitting ACPEL devices based upon ZnS:Tm, Li. Thin Solid Films 283:204–208
9. Hideki K, Yoji Y, Yukiniro M, Yukihisa T (1996) ZnS: Mn thin films electroluminescent devices prepared by metalorganic chemical vapor deposition. J Cryst Growth 169:33–39
10. Goswami N, Sen P (2007) Photoluminescent properties of ZnS nanoparticles prepared by electroexplosion of Zn wires. J Nanoparticle Res 9:513–517
11. Do VN, Tuan NT, Trung DQ, Kien ND, Chien ND, Huy PT (2010) One-dimensional fabricated by oxidizing ZnS nanowires. Mater Lett 64:1650–1652
12. Curie D (1963) *Luminescence of crystals*. Methuen and Co. Ltd, London. (chapters 4 and 5)
13. Cheng YC, Jin CQ, Gao F, Wu XL, Zhong W, Li SH, Chu PK (2009) Raman scattering study of zincblende and wurtzite ZnS. J Appl Phys 106:123505
14. Brafman O, Mitra SS (1968) Raman effect in wurtzite- and zincblende-type ZnS single crystals. Phys Rev 171:931–934
15. Schneider J, Kirby RD (1972) Raman scattering from ZnS polytypes. Phys Rev B 6:1290–1294
16. Siegle H, Kaczmarczyk G, Filippidis L, Litvinchuk AP, Hoffmann A, Thomsen C (1997) Zone-boundary phonons in hexagonal and cubic GaN. Phys Rev B 55:7000–7004

# Part VIII
# Characterization of Intermetallic Materials

This part includes two chapters related to the characterization of intermetallic compounds.

The first chapter study the room-temperature compressive mechanical behavior of intermetallic FeAl40at.% based alloys manufactured by spray atomization and deposition was evaluated. FeAl40at.% alloy doped with boron and reinforcement with alumina particulate showed a diminished in the yield stress consistently with increase of the strain rate. The results showed evidence that alumina particulate plays an important role in the drop of yield stress because it promotes cracks that degrade the mechanical properties.

On the other hand, the second chapter is related to intermetallic compounds with high aluminum content, that are materials from which hydrogen is released due to their susceptibility to hydrogen environment embrittlement. This work take advantage of this phenomenon to generate hydrogen using the Fe2Al5 intermetallic. The compound was activated mechanically to promote and increase the amount of hydrogen released. Fe2Al5 intermetallic compounds with different aqueous solutions were studied, which aqueous solutions were used to increase the pH. The results showed that, as the pH of the solution increases, the amount and rate of hydrogen release is increased.

The main techniques used in the characterization of these materials are:

- Transmission electron microscopy (TEM).
- X-ray diffraction (XRD).
- Scanning electron microscopy (SEM).

# Chapter 20
# Mechanical Properties of Spray-Atomized FeAl40 at.%Al Alloys

**M. Amaya, J.M. Romero, L. Martinez, and R. Pérez**

**Abstract** Three different as-atomized alloys Fe-40 at.%Al, Fe-40Al+0.1 at.%B, and Fe-40Al+0.1 at.%B+10 at.%$Al_2O_3$ were deformed by compression under strain rates from $10^{-4}$ to $10^{-2}$ $s^{-1}$, in order to study room-temperature mechanical properties. In the interval of $10^{-4}$ to $10^{-3}$ $s^{-1}$ strain rates all three alloys diminished their yield stress but in the interval of $10^{-3}$ to $10^{-2}$ $s^{-1}$, the FeAl40 at.% base alloy and the alloy doped with boron raise their yield stress. On the other hand, contrary to the last two alloys the FeAl40 at.% alloy doped with boron and reinforcement with alumina particulate shows a decrease in the yield stress in the interval of the $10^{-3}$ to $10^{-2}$ $s^{-1}$. The yield stress is discussed in terms of the effect of the boron and alumina particles, and the fracture morphology behavior. The main objective of this research is to present a comparative analysis of the three intermetallic materials, evaluating their structure and mechanochemical properties.

**Keywords** Atomized alloys • Iron aluminides • Fracture • Microalloying • Intermetallic

---

M. Amaya
Compañía Mexicana de Exploraciones, Av. M. Escobedo 366, Col. Anzures, Del. Miguel Hidalgo, México D.F., México
e-mail: manama31@hotmail.com

J.M. Romero
Instituto Mexicano del Petróleo, Eje Central Lázaro Cárdenas Norte 152, Col. San Bartolo Atepehuacan, México D.F. 07730, México
e-mail: jromero@imp.mx

L. Martinez
Instituto de Ciencias Físicas, Universidad Nacional Autónoma de México, Ap. Postal 43-8, Cuernavaca, Morelos 62251, México
e-mail: lmg@corrosionyproteccion.com

R. Pérez (✉)
Centro de Física Aplicada y Tecnología Avanzada, Universidad Nacional Autónoma de México, Boulevard Juriquilla 3001, Santiago de Querétaro, Qro. 76230, México
e-mail: ramiro@fata.unam.mx



R. Pérez Campos et al. (eds.), *Materials Characterization*,
DOI 10.1007/978-3-319-15204-2_20

## 20.1 Introduction

During the last two decades several researchers have been reported that the iron aluminides have been shown superior properties compared to the commercial alloys used in high temperature and corrosive environments [1–3]. Nevertheless, their processing and applications have been limiting due to the lack of ductility [4, 5]. In order to improve ductility, novel processing routes such as microalloying, particulate dispersing, fiber reinforcement, and microstructure control have been explored [6, 7]. Extensive studies on the mechanical properties of iron aluminides were realized in the 1980s [8–10]. However, in the 1990s the most useful results about this topic were reported [11–20]. In order to improve mechanical properties, FeAl intermetallic alloys were manufactured using a spray-atomization and deposited metallurgical route. Three different alloys were manufactured, FeAl40 at.% base alloy, FeAl40 at.% base alloy microalloying with boron, and FeAl40 at.% base alloy microalloying with boron and reinforcement with alumina particulate. The effect of the boron and alumina particulate on the stress and the strain mechanical properties at room temperature was studied. In addition, the relation of the mechanical properties and the fracture properties behavior was discussed.

## 20.2 Experimental

Spray-atomization and deposition technique was used to fabricate three different ingots of FeAl intermetallic alloys. The composition of the alloys was FeAl40 at.%, FeAl40 + 0.1 at.%B, and FeAl40 + 0.1 at.%B + 10 at.%$Al_2O_3$. The fabrication details are discussed elsewhere [21]. As-atomized parallelepiped specimens with dimensions of 10 × 5 × 5 mm were sectioned from all the three different ingots using a spark cutting machine following the spray-atomization direction. In order to measure the grain sizes, a face of each specimen was mechanical grinding with wet SiC-paper (grit 600) and subsequently polished using colloidal alumina of 3 and 1 μm. Finally, the polished faces were etching by a solution consisted in 33 % acetic acid, 33 % nitric acid, 1 % fluorhydric acid, and 33 % $H_2O$ (% volume). The intercept method (ASTM-E112) was employed to measure the grain sizes. Compressive mechanical tests were carried out at room temperature using a universal testing machine (Instron 4206), at strain rates (SR) of $10^{-4}$, $10^{-3}$, and $10^{-2}$ $s^{-1}$, according to the standard ASTM D695. Teflon film was used to minimize the friction between specimens and compression pads. Before the compression tests, all parallelepiped specimens were mechanical grinding with wet SiC-paper (grit 600) and subsequently polished using colloidal alumina of 3 and 1 μm. All three different FeAl specimens were strained to failure. The yield stresses were measured at 0.2 % plastic strain (0.2 % offset yield strength). After compression tests, all fracture surfaces of the specimens were examined by a JEOL 400 scanning electron microscopic.

## 20.3 Results and Discussion

The FeAl (FA) and FeAl40+0.1 at.%B (FAB) alloys (Fig. 20.1a, b) showed non-uniform initial grain size distribution as compared to the FeAl40+0.1 at.%B+1 0 at.%$Al_2O_3$ (FAL) alloy (Fig. 20.1c). The presence of alumina particulate seems to induce a grain size refinement and improves the microstructure of the FAL alloy.

The final average grain sizes of the specimens were 250, 160, and 65 μm for the FA, FAB, and FAL alloys, respectively. The true stress–strain curves of all the three as-atomized FeAl base alloys are shown in Fig. 20.2. These curves show that all the three alloys yield continuous and smoothly. Several researches have reported that the fine-grained FeAl alloys yield discontinuously while large-grained yield smoothly [11, 13, 18, 22]. In the FA and FAB alloys, the ultimate strength (US) raises constantly as the strain rate diminishes. However in the FAL alloy, the US increases abruptly in the SR range from $10^{-2}$ to $10^{-3}$ $s^{-1}$. The US remains quasi-constantly in the SR of $10^{-3}$ $s^{-1}$ and $10^{-4}$ $s^{-1}$. Figure 20.3a shows the effect of the SR on the 0.2 % offset yield strength (YS). In the SR range of $10^{-2}$ to $10^{-3}$ $s^{-1}$, the FA and FAB alloys diminished the YS, and then the YS increases up to SR of $10^{-4}$ $s^{-1}$.

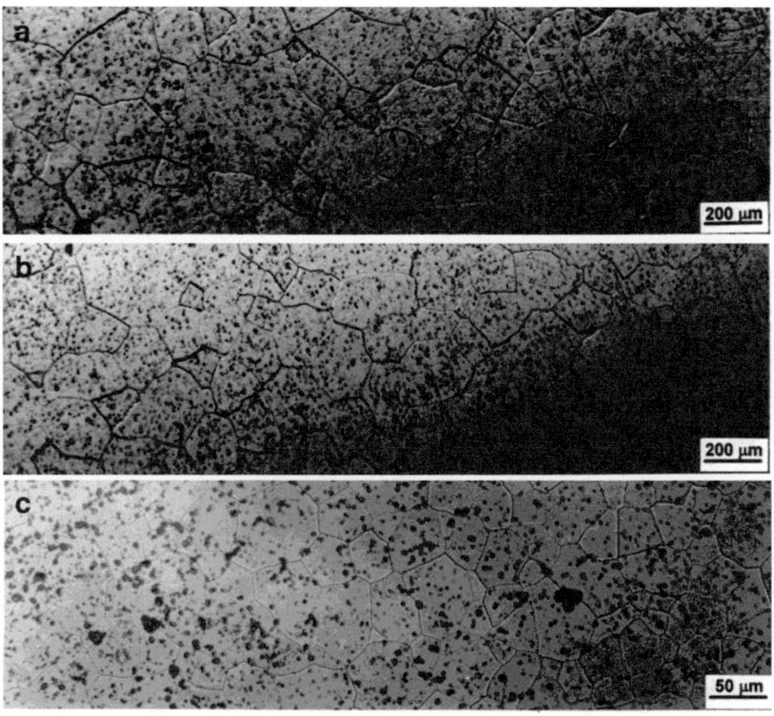


**Fig. 20.1** As-atomized microstructures of the (**a**) FeAl40 at.%, (**b**) FeAl40+0.1 at.%B, and (**c**) FeAl40+0.1 at.%B+10 at.%$Al_2O_3$ intermetallic alloys

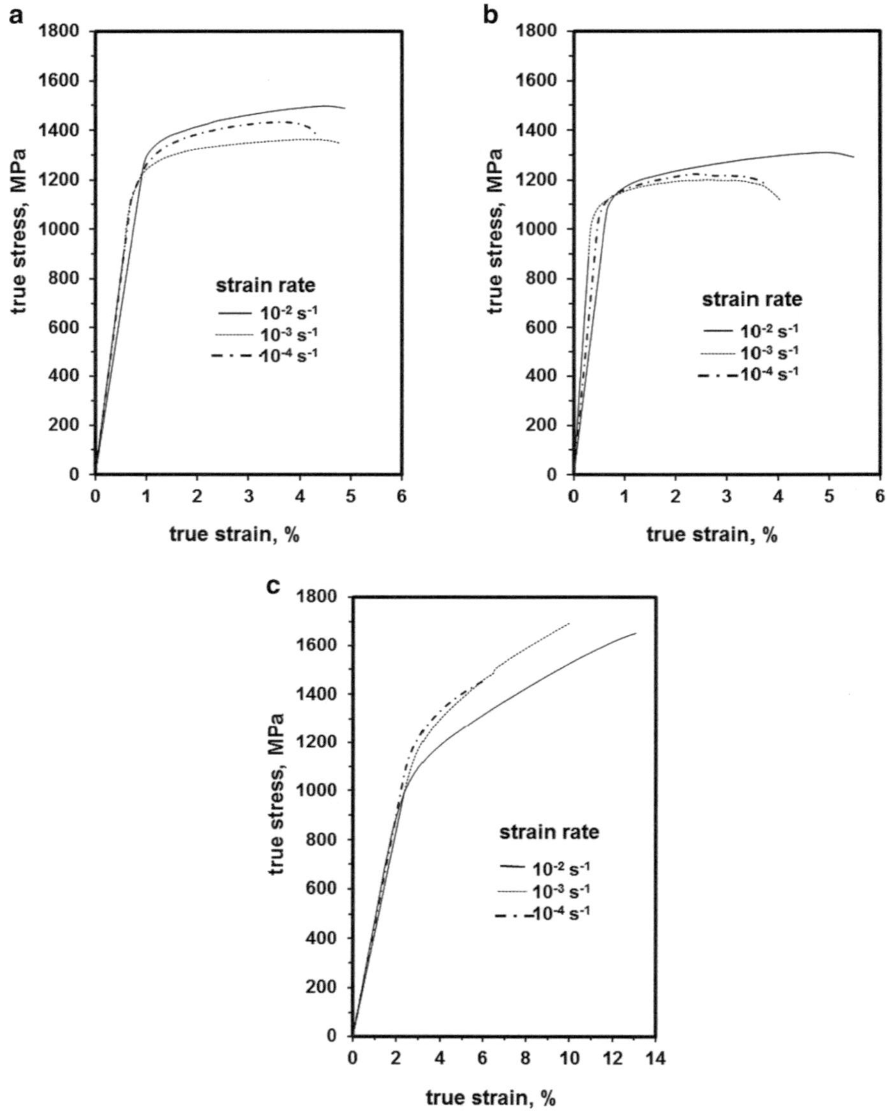


**Fig. 20.2** Stress–strain curves of the (**a**) FeAl40 at.%, (**b**) FeAl40 + 0.1 at.%B, and (**c**) FeAl40 + 0.1 at.%B + 10 at.%$Al_2O_3$ alloys

Also, Fig. 20.3a shows that the boron diminished the YS compared to the YS of the FA base alloy, about 150 MPa.

The effect of the SR on the YS is shown in Fig. 20.3a. Basically, the FA and FAB alloys showed a similar performance. The YS diminished as the strain rate diminished from $10^{-2}$ to $10^{-3}$ $s^{-1}$, then, the YS increased as the strain rate increased from $10^{-3}$ to $10^{-4}$ $s^{-1}$. On the other hand, in the FAL alloy the YS increased constantly, as the strain rate increased from $10^{-2}$ to $10^{-4}$ $s^{-1}$.

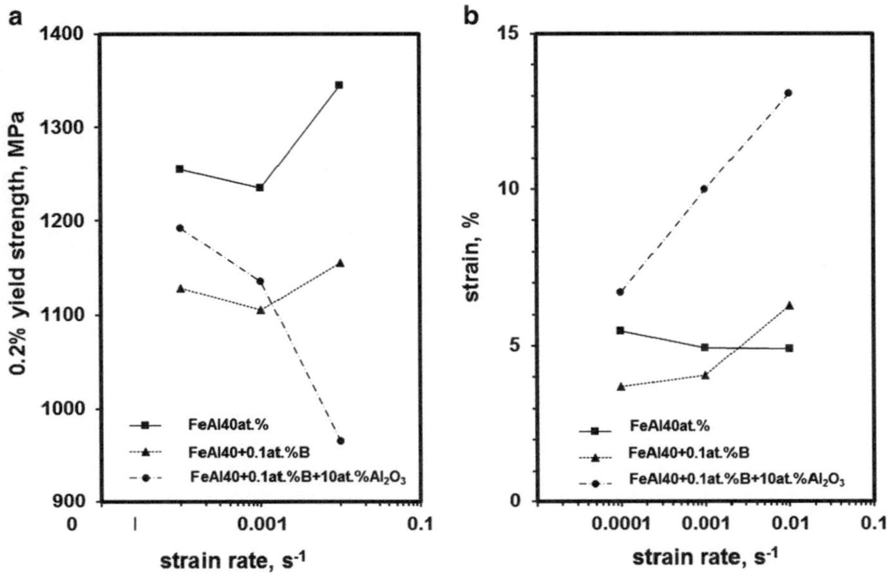


**Fig. 20.3** (**a**) Effect on the strain rate on the 0.2 % yield strength. (**b**) Strain behavior as a function of the strain rate

Figure 20.3b shows the effect of the strain rate on the final strain before the rupture in all the three FeAl alloys. The FA alloy increased the strain smoothly as the strain rate diminished. On the other hand, the FAB and FAL alloys diminished the strain as the strain rate diminished. Compared to the FAB alloy the percentage of strain in the FAL alloy diminished abruptly. The alumina particulate detachment could generate the last effect because these particulates promoted crack niches before the plastic deformation can occur during the deformation.

Figure 20.3 also shows a difference of yield stress between FA and FAB alloys. Other researchers have found that the boron segregates at the grain boundaries improving the cohesion, and suppressing the intergranular fracture [5, 23]. In the present study the pores formed in the microstructure of the FAB can affect the boron distribution in the microstructure. However, no evidence was found of boron distributions. We believe that part of the boron can be segregated in the free surface of the pores, obscuring the beneficial effect on the grain boundaries.

This effect is promoted by the disbondment of the alumina particles, which are the sites for the crack initiation inducing material failures before the plastic deformation. The boron presence affects the deformation nature of the FAB and FAL alloys. The deformation percentage gets lower with the diminution of the strain rate in the FAB and FAL alloys in comparison with FA alloy.

The fracture surfaces obtained in the FeAl alloys are shown in Fig. 20.4. The fracture surface that corresponds to the test conditions of $10^{-3}$ $s^{-1}$ shows more cracks density as compared to the other two strain rates (Fig. 20.4b). The formation of cracks caused by this strain rate can be contributed to drop the yield stress. On the other hand, the drop of yield stress in the FAB alloy can be attributed to

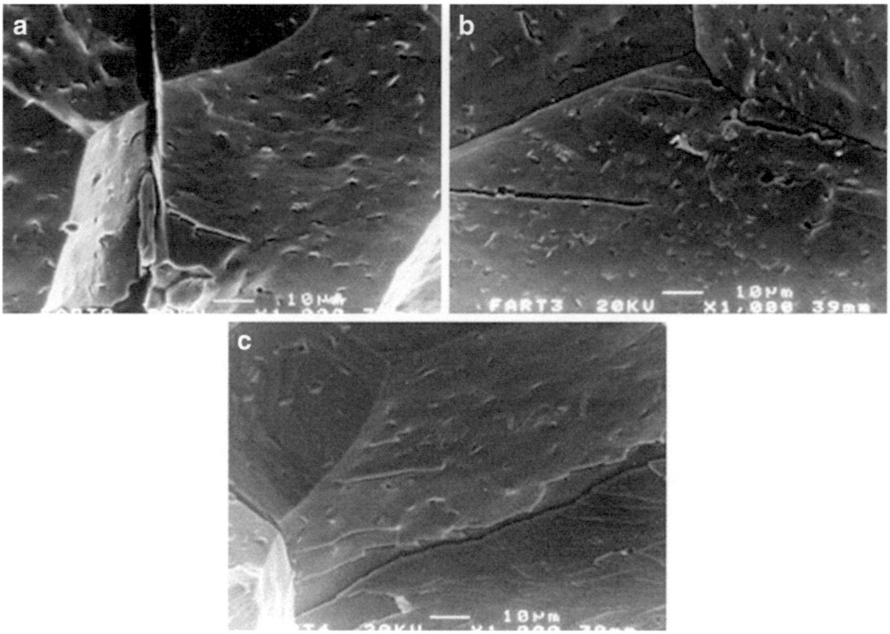

**Fig. 20.4** Fracture morphologies on the FeAl40 at.% alloy obtained at strain rates of (**a**) $10^{-2}$ $s^{-1}$, (**b**) $10^{-3}$ $s^{-1}$, and (**c**) $10^{-4}$ $s^{-1}$

different factors. In this case, the pores seemed to play an important role because they incubate and transmit crack paths (Fig. 20.5b) that contributed to lower microstructural integrity.

In the FAL alloy, the yield stress gets lower values with the increase of the strain rate. This behavior was attributed to the generation of cracks in the grain boundaries which is promoted by the alumina particles. Figure 20.6 shows evidence of cracks and alumina particulates on the grain boundaries.

Figure 20.6 illustrates the evidence of cracks along the grain boundaries, Fig. 20.6a, b, however at the lowest strain rate ($10^{-4}$ $s^{-1}$), less crack density found, Fig. 20.6c. On the other hand, it seems that the strain rate plays an important role in the strength of this alloy as is illustrated in Fig. 20.3.

## 20.4 Conclusions

The room-temperature compressive mechanical behavior of intermetallic FeAl40 at.% based alloys manufactured by spray-atomization and deposition was evaluated. All three alloys studied yield continuously in the strain rate interval of $10^{-4}$ to $10^{-2}$ $s^{-1}$. FeAl40 at.% and FeAl40+0.1 at.%B alloys showed a decrease in the yield stress at strain rates of $10^{-3}$ $s^{-1}$. This behavior was attributed to high density

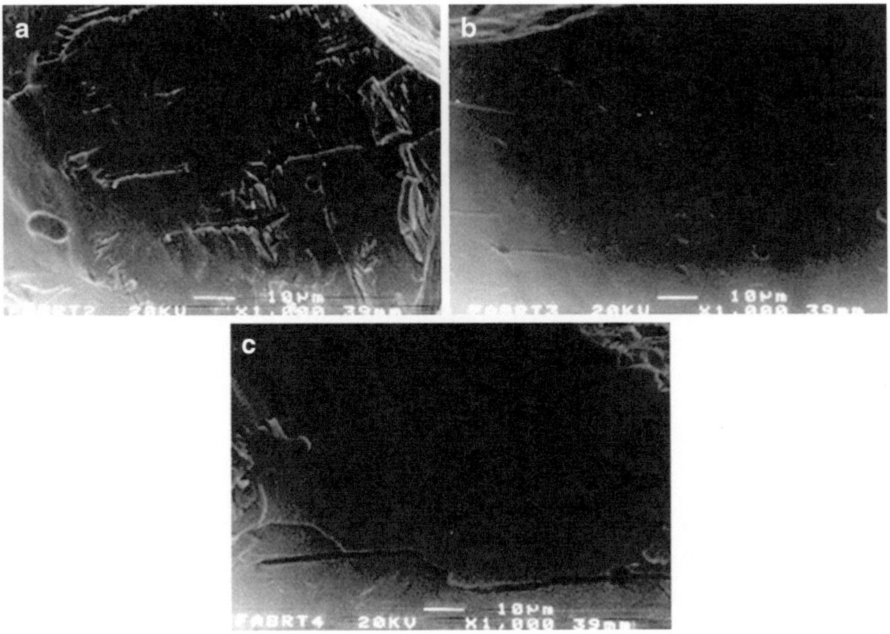

**Fig. 20.5** Fracture morphologies on the FeAl40 + 0.1 at.%B alloy obtained at strain rates of (**a**) $10^{-2}$ $s^{-1}$, (**b**) $10^{-3}$ $s^{-1}$, and (**c**) $10^{-4}$ $s^{-1}$

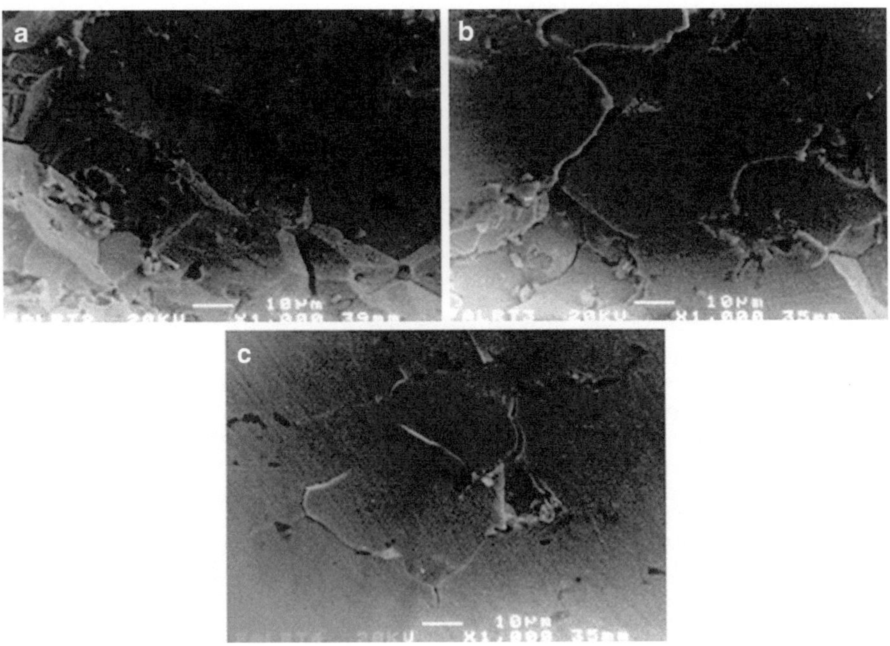

**Fig. 20.6** Fracture morphologies on the FeAl40 + 0.1 at.%B + 10 at.%$Al_2O_3$ alloy obtained at strain rates of (**a**) $10^{-2}$ $s^{-1}$, (**b**) $10^{-3}$ $s^{-1}$, and (**c**) $10^{-4}$ $s^{-1}$

of microcleavage in the FeAl40 at.%, and in the FeAl40+0.1 at.%B it was related to cracks induced by pores that lowers the microstructural integrity. In the case of the FeAl40+0.1 at.%B+10 at.%$Al_2O_3$ alloy the yield stress diminished consistently with increase of the strain rate. Our results show evidence that alumina particulate plays an important role in the drop of yield stress because it promotes cracks that degrade the mechanical properties. The presence of alumina particles significantly inhibits the phenomena of recrystallization and growth of the intermetallic structure. While the structures that are free of particles have a microstructure deformed and also the recrystallization process is observed.

**Acknowledgment** The authors acknowledge the SEM technical support of O. Flores.

## References

1. Yoo MH, Horton JA, Liu CT (1988) Micromechanisms of yield and flow in ordered intermetallic alloys. Acta Metall 36:2935–2946
2. Sikka VK (1994) Processing and applications of iron aluminides. In: Schneibel JH, Crimp MA (eds) Processing, properties and applications of iron aluminides. TMS, Warrandale
3. Amaya M (1999) Mechanical and electrochemical properties of FeAl alloys at high temperatures. PhD Thesis, Universidad Nacional Autonoma de Mexico
4. Liu CT, McKamey CG (1990) Environmental embrittlement – a major cause for low ductility of ordered intermetallics. In: Whang SH, Liu CT, Pope DP, Stiegler JO (eds) High temperature aluminides and intermetallics. TMS, Warrandale
5. Liu CT, George EP (1990) Environmental embrittlement in boron-free and boron-doped FeAl (40 at.% Al) alloys. Scr Metall Mater 24:1285–1290
6. Baker I, Munroe PR (1997) Mechanical properties of FeAl. Int Mater Rev 42:181–205
7. Martinez L, Amaya M, Porcayo-Calderon J, Lavernia EJ (1998) High-temperature electrochemical testing of spray atomized and deposited iron aluminides alloyed with boron and reinforced with alumina particulate. Mater Sci Eng: A 258:306–312
8. Whittenberger JD (1983) Influence of thermomechanical processing on elevated temperature slow plastic flow properties of B2 aluminide Fe-39.8 at.%Al. Mater Sci Eng 57:77–85
9. Baker I, Gaydosh DJ (1987) Flow and fracture of FeAl. Mater Sci Eng 96:147–158
10. Schmidt B, Nagpal P, Baker I (1989) Annealing studies of B2 FeAl. Mater Res Soc Symp Proc 133:755–760
11. Baker I, Nagpal P, Liu F, Munroe PR (1991) The effect of grain size on the yield strength of FeAl and NiAl. Acta Metall Mater 39:1637–1644
12. Pike LM, Liu CT (1991) The effect of boron doping on the Hall–Petch slope of FeAl (40 at.% Al). Scr Metall Mater 25:2757–2761
13. Gaydosh DJ, Draper SL, Noebe RD, Nathal MV (1992) Room temperature flow and fracture of Fe-40 at.%Al alloys. Mater Sci Eng: A 150:7–20
14. Baker I, Klein O, Nelson C, George EP (1994) Effects of boron and grain size on the strain-rate sensitivity of Fe-45Al. Scr Metall Mater 30:863–868
15. Baker I, Xiao H, Klein O, Nelson C, Whittenberger J (1995) The effect of temperature and Fe:Al ratio on the flow and fracture of FeAl. Acta Metall Mater 43:1723–1730
16. Imayev R, Evangelista E, Tassa O, Stobrawa J (1995) Relationship between mechanism of deformation and development of dynamic recrystallization in FeAl intermetallic. Mater Sci Eng: A 202:128–133
17. Scheff SA, Stout JJ, Crimp MA (1995) Effect of extrusion texture on the compressive behavior of B2 FeAl alloys. Scr Metall Mater 32:975–980

18. Morris DG, Gunther S (1996) Strength and ductility of Fe40Al alloy prepared by mechanical alloying. Mater Sci Eng: A 208:7–19
19. Baker I, Yang Y (1997) On the yield stress anomaly in stoichiometric FeAl. Mater Sci Eng: A 239–240:109–117
20. Cohron JW, Lin Y, Zee RH, George EP (1998) Room-temperature mechanical behavior of FeAl: effects of stoichiometry, environment, and boron addition. Acta Mater 46:6245–6256
21. Martinez L, Flores O, Amaya M, Duncan A, Viswanathan S, Lawrynowics D, Lavernia EJ (1997) The role of alumina particulate in microstructural and forging properties of spray atomized and deposited Fe–Al ordered intermetallic compounds. J Mater Synth Process 5:65–76
22. Briguet C, Morris DG (1997) Deformation mechanics in a mechanically alloyed Fe-40Al alloy and the influence of recrystallization and ageing heat treatments. Acta Mater 45:4939–4951
23. Crimp MA, Vedula KM (1986) Effect of boron on the tensile properties of B2 FeAl. Mater Sci Eng 78:193–200

# Chapter 21
# Structural Characterization of $Fe_2Al_5$ Intermetallic Compound After Reaction with Water to Release Hydrogen

**J. Luis López-Miranda, R. Esparza, and G. Rosas**

**Abstract** Intermetallic compounds with high aluminum content are materials from which hydrogen is released due to their susceptibility to hydrogen environment embrittlement. Therefore, the present work take advantage of this phenomenon to generate hydrogen using the $Fe_2Al_5$ intermetallic. The compound was activated mechanically to promote and increase the amount of hydrogen released. Furthermore, NaOH aqueous solutions were used to increase the pH and thereby increase both the rate and amount of hydrogen. The results showed that, as the pH of the solution increases, the amount and rate of hydrogen release is increased. Finally, characterization of solid by-products of the reaction was performed by X-ray diffraction (XRD), transmission electron microscopy (TEM), and scanning electron microscopy (SEM). All cases showed that the bayerite phase is the major product of the hydrogen generation reaction.

**Keywords** Intermetallic compounds • Hydrogen generation • Aqueous solutions • pH • Structural characterization

## 21.1 Introduction

In recent years, the world has faced an energy crisis due to the depletion of resources and increasing environmental problems [1–3]. This creates the need for the use of alternative fuels. One of these is the hydrogen which is the best clean fuel that

---

J.L. López-Miranda (✉) • G. Rosas
Instituto de Investigaciones Metalúrgicas, UMSNH, edificio U, Ciudad Universitaria, Morelia Michoacán 58000, México
e-mail: lopezfim@gmail.com; grtrejo07@yahoo.com.mx

R. Esparza
Centro de Física Aplicada y Tecnología Avanzada, Universidad Nacional Autónoma de México, Boulevard Juriquilla 3001, Santiago de Querétaro, Qro. 76230, México
e-mail: resparza@fata.unam.mx


R. Pérez Campos et al. (eds.), *Materials Characterization*,
DOI 10.1007/978-3-319-15204-2_21

produces energy through a fuel cell by converting chemical energy into electricity with a high yield under relatively simple conditions. However, solving the problem of on-site storage of this gas and the development of effective methods for their production are prerequisites to fulfill these expectations. In order to solve this problem, there are various proposed methods for the generation of hydrogen in situ. One of these is the hydrolysis of water from the reaction between aluminum and water according to Eq. (21.1):

$$2Al + 6H_2O \rightarrow 2Al(OH)_3 + 3H_2 \quad \Delta G = -427\ \text{kJ} \tag{21.1}$$

However, pure aluminum cannot react with water due to a thin oxide layer overlying, so it is necessary to carry out an activation process [3–5]. The main activation methods are: the aluminum alloyed with certain metals such as Ga, In, and Hg [5– 7], the use of additives [8, 9], and the use of alkaline solutions using substances such as CaO, NaOH, and KOH [10, 11]. Another method of activation is the mechanical milling, which promotes the hydrogen generation reaction. This is because the crystal and particle refinement, the point defects and the release of fresh surface, clean of surface oxide [12–14]. Moreover, it is known that the intermetallic compounds are susceptible to environmental embrittlement, that is, they are materials having the ability to react with atmospheric moisture to release hydrogen, pulverizing the alloy in a few days [15]. In this research a novel hydrogen production process was investigated in order to find a clean and safe energy generation method. Therefore, the results of the hydrogen generation by the reaction between aluminum and water are shown. This was performed using the $Fe_2Al_5$ intermetallic compound activated by mechanical milling and whose reaction for hydrogen release was taken out using NaOH and CaO aqueous solutions.

## 21.2 Experimental

The starting materials were high purity elements, aluminum and iron (99.97 % purity). The $Fe_2Al_5$ intermetallic alloy was prepared by conventional casting using silicon carbide crucibles. Based on the equilibrium diagram and assuming the atomic formula ($Fe_2Al_5$), the weight percentage was determined for 1 kg of intermetallic. These percentages were 54.71 % Al and 45.29 % Fe. The cast was cooled and solidified into a metal mold at ambient conditions. The ingot alloy broke into fine particles due to the hydrogen embrittlement susceptibility of this kind of materials. Then, the intermetallic powder was mechanically activated in a high energy ball mill SPEX 8000M. The milling media were hardened steel vials and 2 steel balls of ½″ diameter. The only parameter varied in the mechanical milling was the time corresponding to 1, 5, and 10 h. The ball to powder ratio weight was constant in all experiments with a value of 5.6. To carry out the reaction of hydrogen generation, first was evaluated the mechanical activation time. Subsequently the pH of the solutions was varied, with some substances, CaO powder (Alfa Aesar) and NaOH

(Alfa Aesar) in flake form. Each of the substances (CaO and NaOH) was dissolved in deionized water at various concentrations in order to have different alkaline pH values. On the other hand, the hydrogen evolution was analyzed by means of a displacement method which consists of inverted burette water filled. The container with the intermetallic and the NaOH solution was connected to burette. So, the measurement of hydrogen gas generated was conducted for the water displacement. After the hydrogen generation, the samples were dried at ambient conditions to obtain the solid products. Subsequently, the solid samples were characterized to determine the by-products generated and the hydrogen generation mechanism. The solid products characterization was carried out by X-ray diffraction (Siemens D5000 X-ray diffractometer), scanning electron microscopy (JEOL JSM 6400), and transmission electron microscopy (Phillips Tecnai F20).

## 21.3 Results and Discussion

Figure 21.1a shows the XRD pattern of the as-cast sample with nominal composition of $Fe_2Al_5$. From the pattern, it is observed a mixture of intermetallic phases which consist in the $FeAl_2$-triclinic phase (JCPDS 00-033-0019) with lattice parameters of $a=7.609$ Å, $b=16.91$ Å, $c=4.869$ Å, ($\alpha=89.49°$, $\beta=122.65°$, $\gamma=90.54°$) and $Fe_2Al_5$-orthorhombic phase (JCPDS 00-047-1435) with cell parameters of $a=7.648$ Å, $b=6.413$ Å, $c=4.216$ Å. It is well known that it is difficult to obtain an intermetallic

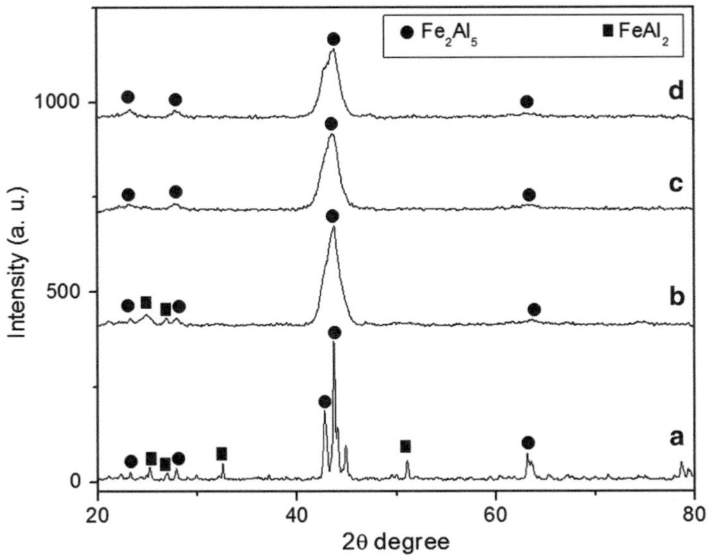


**Fig. 21.1** XRD analysis showing the milling time effect on the $Fe_2Al_5$ intermetallic structure: (**a**) as-cast, (**b**) 1, (**c**) 5 and (**d**) 10 h of milling

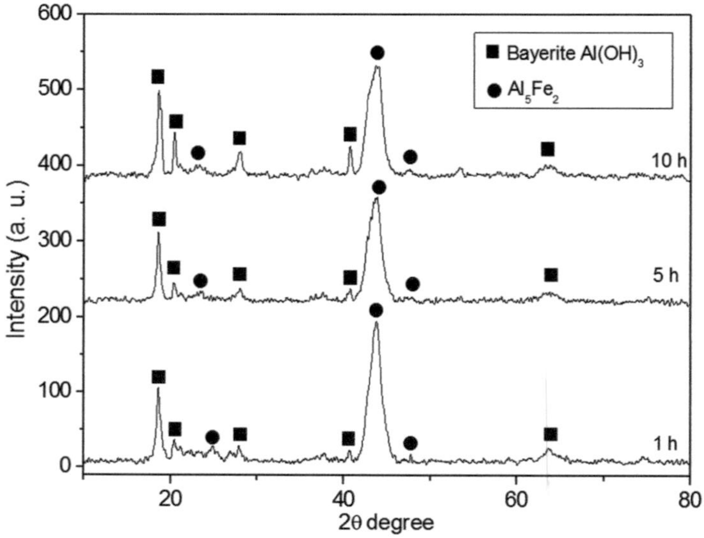


**Fig. 21.2** XRD analysis of by-products of hydrogen generation using $Fe_2Al_5$ activated at different times

single phase in the conventional casting methods. This is due to the great chemical phase segregation existing during solidification of the alloy. This is also attributed to narrow compositional ranges within the Fe–Al equilibrium phase diagram of these intermetallic phases. Figure 21.1b–d shows XRD patterns of the milled samples for 1, 5, and 10 h of milling, respectively. XRD peaks of $Fe_2Al_5$ intermetallic phase increased while peaks related of $FeAl_2$ phase decreased as the milling time increased. According to these results, the milling treatment at different times of the intermetallic sample seems to behave as a homogenizing treatment. Likewise, a broadening in the diffraction peaks occurs due to the crystal size reduction, disordering and increase of internal stresses in the lattice when the milling time is increased. This has been demonstrated in other researches where several Al–Fe system intermetallic compounds were subjected to high energy ball-milling [16, 17].

Figure 21.2 shows the crystalline phases which are obtained when the above milled powders are reacted with an aqueous solution of pH = 12 using NaOH additions. This is in order to analyze whether the milled powders can react easily with the solution to release hydrogen. It is clearly evident from the three diffraction patterns in Fig. 21.2 the presence of the monoclinic bayerite phase ($Al(OH_3)$) and starting intermetallic phase orthorhombic-type ($Al_5Fe_2$). The presence of bayerite phase is an indication that the reaction between the aluminum of the intermetallic phase and the aqueous solution takes place to release hydrogen. In all XRD patterns, similar peaks intensities are also observed. With the above it is concluded that among 1 and 10 h of milling time, there are not an appreciable effect on the hydrogen generation reaction. This is due to oxidation of aluminum increases with the milling time, which has been determined in other works [6, 18, 19]. Thus, all reactions to assess

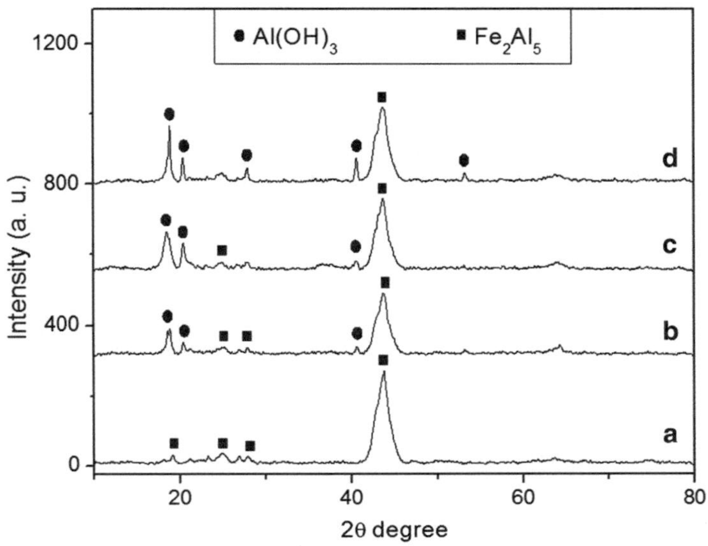


**Fig. 21.3** XRD analysis of hydrogen generation reaction by-products using CaO solutions at different pH values: (**a**) 7, (**b**) 10, (**c**) 12 and (**d**) 13

the amount of hydrogen released are presented hereafter using milled powders for 1 h. These results are similar to those previously obtained by Chen, who used different milling times obtaining the best results for the sample milling for 2 h. Prolonged ball-milling decreased the rate and amount of hydrogen released [18].

Figure 21.3 shows the XRD patterns obtained after exposing the intermetallic milled powders at different aqueous solutions of pH from 7 to 13. The aqueous solution of pH = 7 corresponds to the reaction performed with deionized water, while for reactions performed in solutions at higher pH values, different additions of CaO were used. A slight increase in the intensities of the bayerite-phase diffraction peaks directly related to the amount of hydrogen released as the alkalinity of the reaction solution is increased up to pH = 13 is observed. This is according to Pourbaix diagrams, whereas the pH of the solution is increased the field of the hydroxide phase is more stable. Previous research has also shown that the use of an alkaline solution can reduce the passivation of the aluminum surface and increasing the rate of hydrogen released [10, 20]. The increases of pH value with CaO is about to $Ca(OH)_2$ formation according to the following reaction:

$$CaO + H_2O \rightarrow Ca(OH)_2 \tag{21.2}$$

Therefore, the hydrogen gas generated depends strongly on the CaO content. In some researches, 100 % efficient reactions have been achieved with CaO additions, i.e., all aluminum is consumed for hydrogen generation [11, 21]. However, as can be seen in Fig. 21.3, in this work the intermetallic compound is present for the three

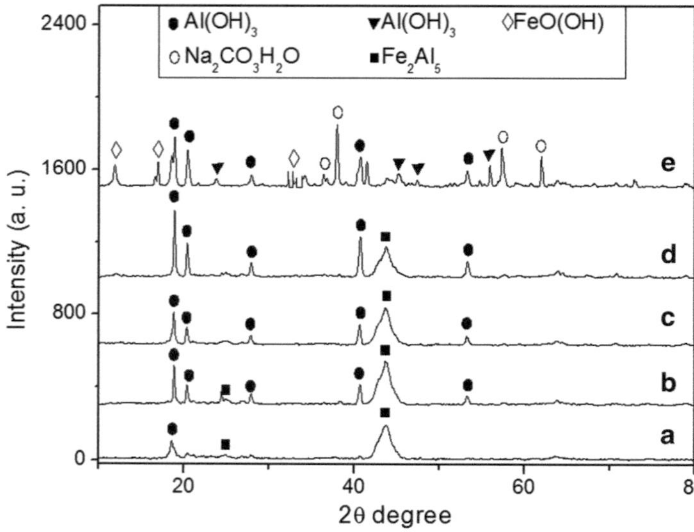


**Fig. 21.4** XRD analysis of hydrogen generation reaction by-products using (**a**) deionized water and NaOH solutions at different pH values: (**b**) 10, (**c**) 12, (**d**) 13 and (**e**) 14

alkaline solutions evaluated. Consequently, high efficient reactions to hydrogen release were not achieved.

In order to compare the results obtained with the use of other additives, some experiments were performed by adding NaOH to adjusting in the reaction solution the pH at different values (10, 11, 12, and 14). Figure 21.4 shows the XRD patterns of the obtained by-products. In order to compare the results, the pattern corresponding to the sample with pH = 7 is presented. From the XRD patterns, as pH increases the intensities of bayerite peaks are increased, indicating an increased tendency of aluminum to react with water and release hydrogen. However, when the reaction is carried out at pH = 14, the process approaches its maximum performance, that is, all aluminum contained in the intermetallic alloy ($Fe_2Al_5$) reacts with water. This is deduced from the presence of iron oxyhydroxide phases (FeO(OH)) that are formed according to the corresponding XRD pattern. From the same XRD patterns (pH = 14) a phase of hydrated sodium carbonate ($Na_2CO_3H_2O$) is also observed. This phase is formed when the dry NaOH reacts with $CO_2$ environment. According to these results, we conclude that the use of NaOH additive is better than CaO. These results agree with several investigations performed by other researchers [10, 22]. In this case, the hydrogen is released according to the following mechanism:

$$2\mathrm{Al} + 6\mathrm{H_2O} + 2\mathrm{NaOH} \rightarrow 2\mathrm{NaAl}\left(\mathrm{OH}\right)_4 + 3\mathrm{H_2} \quad \Delta G = -476\ \mathrm{kJ} \tag{21.3}$$

$$\mathrm{NaAl}\left(\mathrm{OH}\right)_4 \rightarrow \mathrm{NaOH} + \mathrm{Al}\left(\mathrm{OH}\right)_3 \tag{21.4}$$

where the NaOH is regenerated from $NaAl(OH)_4$ compound formed on Eq. (21.3). According to Eq. (21.3), the hydrogen generation depends on alkali concentration. Therefore the overall reaction of hydrogen generation is similar to Eq. (21.1). NaOH

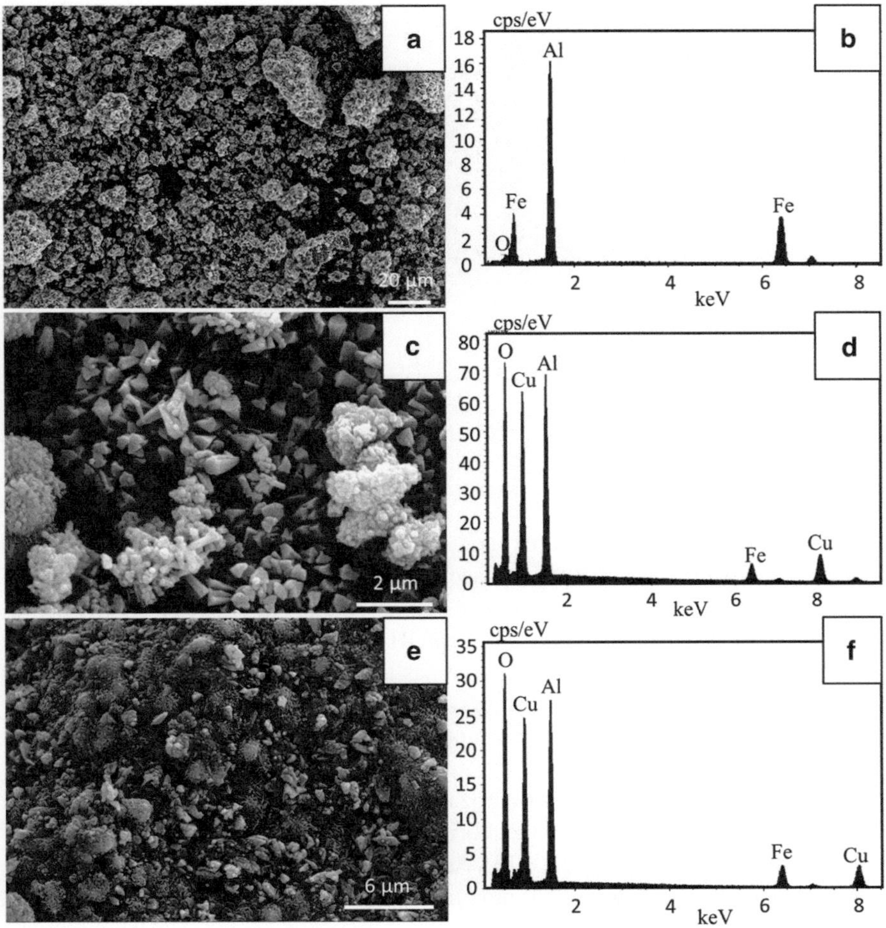


**Fig. 21.5** SEM images showing the typical morphology of: (**a**) intermetallic compound and hydrogen generation by-products using (**c**) CaO, (**e**) NaOH solutions at pH=13 and (**b**), (**d**), (**f**) their respective EDS

has been widely used for these purposes. Skrovan et al. obtained higher hydrogen release when the NaOH concentration was increased [23].

Figure 21.5 shows a SEM analysis. Figure 21.5a shows the morphology of $Fe_2Al_5$ intermetallic compound before to subject the reaction of hydrogen generation. Moreover, the EDS analysis is shown in Fig. 21.5b. As can be seen only Fe and Al signals are present. Therefore any contamination element is not present in the intermetallic. Figure 21.5c, d shows the SEM images and the corresponding EDS analysis of the $Fe_2Al_5$ intermetallic after the reaction with CaO additions (pH=13). The SEM micrograph shows the peculiar morphology of the elongated crystals corresponding to bayerite phase. The presence of these crystals is compelling evidence that the reaction between aluminum and water proceeds to generate hydrogen gas. This morphology is typical of aluminum hydroxides and has been showed in several

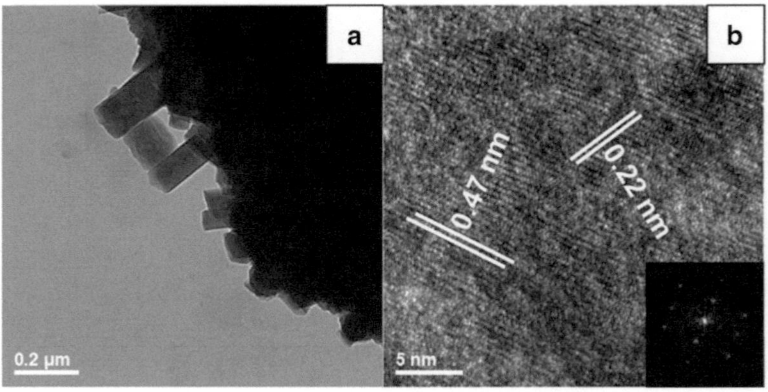


**Fig. 21.6** (**a**) TEM image showing the bayerite crystal size and (**b**) HRTEM image showing lattice planes of bayerite phase

researches [13, 20, 24, 25]. The oxygen element present in the EDS spectrum also indicates the formation of $Al(OH)_3$ phase which is consistent with the SEM images. In the same way, the SEM image and the corresponding EDS spectrum corresponding to the intermetallic sample with NaOH additions (pH = 13) are shown in Fig. 21.5e, f. As can be seen, the sample has the same bayerite crystals, as can be previously determined by XRD patterns. However, the crystals are small and abundant indicating greater crystal nucleation rate compared to the sample with CaO additions.

To complete the structural characterization, the reacted samples were also studied by TEM. Figure 21.6a shows a typical bright-field TEM-image of intermetallic samples with additions of NaOH (pH = 13). As can be seen, the images show the bayerite small crystals with size about 500 nm, confirming that the crystal nucleation rates, and therefore, the hydrogen released rates are better in the NaOH additions. Figure 21.6b shows a HRTEM image of the crystals showing lattice planes for bayerite phase with lattice spacings of 0.47 and 0.22 nm, which correspond to the (001) and (131) lattice planes, respectively.

## 21.4 Conclusions

In this research, we studied the structural evolution of the $Fe_2Al_5$ intermetallic powders after its reaction with aqueous solutions at different pH. When the pH is increased from 10 to 14, the bayerite phase is increased regardless of addition type made (CaO or NaOH), leading to a proportional increase in the hydrogen generation according to reaction (21.1). In the case of solutions with NaOH and for pH = 14, the presence of iron oxyhydroxide phase (FeO(OH)) suggests that all Al in the alloy has reacted with water to form hydrogen. It was observed by SEM and TEM observations that the bayerite phase crystals are small and abundant with NaOH addition (pH = 13) which indicates greater crystal nucleation rates and higher amounts of hydrogen generation. Furthermore, it was observed that there is no appreciable effect on hydrogen generation when milling time varied from 1 to 10 h. Therefore, this material is an excellent alternative to energy generation from hydrogen which can be employed in mobile and stationary applications.

## References

1. Barnwal BK, Sharma MP (2005) Prospects of biodiesel production from vegetable oils in India. Renew Sust Energy Rev 9:363–378
2. Balat M (2008) Potential importance of hydrogen as a future solution to environmental and transportation problems. Int J Hydrogen Energy 33:4013–4029
3. Gai WZ, Liu WH, Deng ZY, Zhou JG (2012) Reaction of Al powder with water for hydrogen generation under ambient condition. Int J Hydrogen Energy 37:13132–13140
4. Xia-Ni H, Chun-Ju L, Yang W, Hang-Yan S, Da C, Yue-Xiang H (2012) Hydrogen generation from hydrolysis of aluminum/graphite composites with a core–shell structure. Int J Hydrogen Energy 37:7457–7463
5. Ilyukhina AV, Ilyukhin AS, Shkolnikov EI (2012) Hydrogen generation from water by means of activated aluminum. Int J Hydrogen Energy 37:16382–16387
6. Fan M, Sun L, Xu F (2010) Study of the controllable reactivity of aluminum alloys and their promising application for hydrogen generation. Energy Convers Manage 51:594–599
7. Jeffrey TZ, Jerry MW, Robert AK, Go C (2011) Liquid phase-enabled reaction of Al–Ga and Al–Ga–In–Sn alloys with water. Int J Hydrogen Energy 36:5271–5279
8. Huihu W, Ying C, Shijie D, Zhifeng L, Qingbiao Z, Ping L, Zhixiong X (2013) Investigation on hydrogen production using multicomponent aluminum alloys at mild conditions and its mechanism. Int J Hydrogen Energy 38:1236–1243
9. Skrovan J, Alfantazi A, Troczynski T (2011) Hydrogen generation by accelerating aluminum corrosion in water with alumina. World Acad Sci Eng Technol 79:325–330
10. Macanás J, Soler L, Candela AM, Muñoz M, Casado J (2011) Hydrogen generation by aluminum corrosion in aqueous alkaline solutions of inorganic promoters: the AlHidrox process. Energy 36:2493–2501
11. Pudukudy M, Yaakob Z, Narayanan B, Ramakrishnan R, Viswanathan S (2012) Hydrogen production from sea water using waste aluminium and calcium oxide. Int J Hydrogen Energy 37:7451–7456
12. Razavi-Tousi SS, Nematollahi GA, Ebadzadeh T, Szpunar JA (2013) Modifying aluminum–water reaction to generate nano-sized aluminum hydroxide particles beside hydrogen. Powder Technol 241:166–173
13. Hsin-Te T, To-Ying L, Yu-Kuang C, Hong-Wen W, Guozhong C (2012) Effect of $Al(OH)_3$ on the hydrogen generation of aluminum–water system. J Power Sources 219:16–21
14. Razavi-Tousi SS, Szpunar JA (2013) Effect of structural evolution of aluminum powder during ball milling on hydrogen generation in aluminum–water reaction. Int J Hydrogen Energy 38:795–806
15. Pouillier E, Gourgues AF, Tanguy D, Busso EP (2012) A study of intergranular fracture in an aluminium alloy due to hydrogen embrittlement. Int J Plasticity 34:139–153
16. López-Miranda JL, Romero-Romero JR, Esparza R, Rosas G (2013) Structural evolution of $FeAl_3$ intermetallic during high-energy ball-milling. Mater Sci Forum 755:133–138

17. Romero-Romero JR, López-Miranda JL, Esparza R, Espinosa-Medina MA, Rosas G (2014) Phase evaluation and its hydrogen correlation of the $FeAl_3$ and $FeAl_2$ intermetallic alloys during mechanical ball-milling with water. Mater Sci Forum 793:143–149
18. Chen X, Zhao Z, Liu X, Hao M, Chen A, Tang Z (2014) Hydrogen generation by the hydrolysis reaction of ball-milled aluminium–lithium alloys. J Power Sources 254:345–352
19. Liu Y, Wang X, Liu H, Dong Z, Li S, Ge H, Yan M (2014) Improved hydrogen generation from the hydrolysis of aluminum ball milled with hydride. Energy 72:421–426
20. Jia Y, Shen J, Meng H, Dong Y, Chai Y, Wang N (2014) Hydrogen generation using a ball-milled Al/Ni/NaCl mixture. J Alloys Compd 588:259–264
21. Chen X, Zhao Z, Hao M, Wang D (2013) Research of hydrogen generation by the reaction of Al-based materials with water. J Power Sources 222:188–195
22. Ma G-L, Dai H-B, Zhuang D-W, Xia H-J, Wang P (2012) Controlled hydrogen generation by reaction of aluminum/sodium hydroxide/sodium stannate solid mixture with water. Int J Hydrogen Energy 37:5811–5816
23. Skrovan J, Alfantazi A, Troczynski T (2009) Enhancing aluminum corrosion in water. J Appl Electrochem 39:1695–1702
24. Wang H-W, Chung H-W, Teng H-T, Cao G (2011) Generation of hydrogen from aluminum and water-Effect of metal oxide nanocrystals and water quality. Int J Hydrogen Energy 36:15136–15144
25. Deng Z-Y, Zhu L-L, Tang Y-B (2010) Role of particle sizes in hydrogen generation by the reaction of Al with water. J Am Ceram Soc 93:2998–3001

# Author Index





R. Pérez Campos et al. (eds.), *Materials Characterization*,
DOI 10.1007/978-3-319-15204-2

# Subject Index





R. Pérez Campos et al. (eds.), *Materials Characterization*,
DOI 10.1007/978-3-319-15204-2

MIX
Papier aus verantwortungsvollen Quellen
Paper from responsible sources
FSC® C105338

If you have any concerns about our products,
you can contact us on
**ProductSafety@springernature.com**

In case Publisher is established outside the EU,
the EU authorized representative is:
**Springer Nature Customer Service Center GmbH**
**Europaplatz 3, 69115 Heidelberg, Germany**

Printed by Libri Plureos GmbH
in Hamburg, Germany